Sarwar Uddin Ahmed
Keinosuke Gotoh

Cost-Benefit Analysis of Environmental Goods by Applying the Contingent Valuation Method

Some Japanese Case Studies

Sarwar Uddin Ahmed
Keinosuke Gotoh

Cost-Benefit Analysis of Environmental Goods by Applying the Contingent Valuation Method

Some Japanese Case Studies

With 68 Figures

Springer

Sarwar Uddin Ahmed
Department of Civil Engineering
Faculty of Engineering
Nagasaki University
1-14 Bunkyo-cho, Nagasaki 852-8521, Japan

Keinosuke Gotoh
Graduate School of Science and Technology
Nagasaki University
1-14 Bunkyo-cho, Nagasaki 852-8521, Japan

Library of Congress Control Number: 2005937066

ISBN-10 4-431-28949-6 Springer-Verlag Tokyo Berlin Heidelberg New York
ISBN-13 978-4-431-28949-4 Springer-Verlag Tokyo Berlin Heidelberg New York

Springer is a part of Springer Science+Business Media
springeronline.com

Printed in Japan

Typesetting: Camera-ready by the authors
Printing and binding: Hicom, Japan

Printed on acid-free paper

Sarwar Ahmed dedicates this book to his parents Manik and Pushpa and his wife Sonia and son Ryan. *Tomader jonnoi amar eto kosto.*

Keinosuke Gotoh dedicates this book to his wife Ikuko and his sons Takehiro, Kensuke and Masatomo. *Kansha.*

Preface

We want to preserve our environment for future generations and make it as safe and clean as possible. However, sometimes development activities affecting the environment are inevitable for various reasons, such as protection against natural disasters, making more land available for housing or farming, building infrastructure facilities, and even redeveloping an environmentally damaged area. Thus the decisions to initiate development projects involving direct or indirect use of environmental goods are very important and must be evaluated carefully, as nothing comes for free. If we want to get something, then we have to forgo something. Before making any decision regarding the use of environmental goods, we need to compare the benefits of preserving them compared with the opportunity costs or forgone benefits for alternative uses, i.e., a cost–benefit analysis (CBA) should be conducted.

This kind of analysis is crucial in the case of Japan, where the environment was partly ignored during the high economic growth era (1955–1973) and where huge amounts of funds still are spent each year for public and private projects that directly or indirectly affect the environment. Hence, it is interesting to investigate two important questions: First, what types of CBA methods are used to justify the initiation of such projects? And second, how are socioenvironmental aspects considered in the analyses? To investigate these questions in this study, we selected the Isahaya Bay Wetland and the theme park Huis Ten Bosch as two investigative cases. We have chosen a total of three cases from divergent viewpoints—public versus private, ecological destruction versus ecological restoration, for instance—to see the changes in valuation concepts and thus CBA analysis.

The main objective of this book is to suggest new directions for assisting cost–benefit analysis of environmental goods and services by linking a potential behavior-contingent valuation method with another multidisciplinary technique, namely, remote sensing. The more specific objectives are: (a) to estimate the value of public and private environmental amenities by applying the contingent valuation method (CVM); (b) to examine the nature and type of cost–benefit analysis (CBA) methods undertaken in Japan concerning environmental amenities; (c) to suggest ways to incorporate socioenvironmental factors into the cost–benefit analysis; (d) to develop models to trace the factors influencing people's willingness to pay for the environmental goods under study and the direction of the impact; and (e) to find ways to make possible environmental valuation by integrating the CVM with various multidisciplinary techniques.

As environmental valuation methodology, CVM was used, and as a data source, mail surveys were conducted followed by extensive studies. In addition, the environmental valuation was supplemented by estimating temporal land-cover change conducted with the help of satellite image analysis from LANDSAT5.

This book is organized into six parts. Part One discusses the introduction and theoretical background of the study. It contains two chapters: Chapter 1 discusses the introduction, theory, and literature behind environmental valuation; the basic concept behind cost–benefit analysis; and includes a summary of various commonly used environmental valuation methods. Chapter 2 thoroughly elaborates the contingent valuation method, which is used in this study along with other environmental valuation methods; its historical development, merits, and demerits; and points that must be given special attention. Part Two, comprising chapters 3 through 5, presents the environmental valuation of the first case study, the Isahaya Bay Wetland (IBW). Chapter 3 provides an overview of the IBW and the Isahaya Bay Reclamation Project (IBRP) and the problems to be addressed. Chapter 4 summarizes the environmental valuation results derived by using CVM. Finally, Chapter 5 recalculates the cost–benefit analysis of the IBRP by applying CVM. Part Three, consisting of Chapter 6, discusses the second case study, regarding the valuation of public parks. Part Four presents the third case study, the environmental valuation of Huis Ten Bosch (HTB). It comprises three chapters, 7 through 9. Chapter 7 provides a background of the theme park HTB and its various initiatives to restore an industrial wasteland. Chapter 8 discusses the environmental valuation results estimated by applying CVM. Finally, Chapter 9 recalculates the environmental cost–benefit analysis of HTB by applying CVM.

Part Five constitutes a unique feature of this book, where by using the database of the three case studies, viz., IBW, public parks, and HTB, supplementary analysis provides some new guidelines for conducting cost–benefit analysis of environmental goods. This part contains four chapters, 10 through 13. First, Chapter 10, using a multivariate model, shows the relationship between distance and willingness to pay. Chapter 11 presents the importance of free comments included at the end of the CVM questionnaire by another multivariate model and graphical summary. Chapter 12 explores the possibility of obtaining a wider view of environmental valuation by integrating CVM and remote sensing. Chapter 13 shows the impact of question format on the willingness to pay. Finally, Part Six, consisting of Chapter 14, summarizes the findings of the case studies and provides policy recommendations for conducting future environmental valuation studies.

This book is the outcome of the efforts of many people: students, reviewers, and colleagues. We are indebted to Professor Shigeru Uchida, Professor Takafumi Yoshida, Dr. Arifa Nazneen, and Dr. A.M. Abbas of Nagasaki University for their thoughtful reviews and suggestions. Special thanks go to Toshihiko Imaoka of Nagasaki University for proofreading and for reducing our workload while we were busy writing the manuscript of this book. For the data input and analysis, we would like to thank all the students of the Environment, Urban Planning, and Remote Sensing Lab, Nagasaki University. We also thank the staff members of Springer-Verlag, Tokyo for their wholehearted support for the publication of this book.

Finally, we are deeply grateful to our family members for their patience, understanding, and encouragement for completion of this book.

Sarwar U. Ahmed
Nagasaki University

Keinosuke Gotoh
Nagasaki University

Contents

List of Abbreviations

BC	Benefit-cost
BG	Bidding Game
CBA	Cost-benefit Analysis
CCM	Contingent Choice Method
CERCLA	Comprehensive Environmental Response, Compensation and Liability Act
CVM	Contingent Valuation Method
DC	Dichotomous Choice
DIS	Disaster Insurance System
EVOS	Exxon Valdez Oil Spill Trustee Council
FAO	Food and Agriculture Organization
HPM	Hedonic Pricing Method
HTB	Huis Ten Bosch
IBRP	Isahaya Bay Reclamation Project
IBW	Isahaya Bay Wetland
JWAN	Japan Wetlands Action Network
LR	Logistic Regression
MAFF	Ministry of Agriculture Forestry and Fisheries
MDA	Multivariate Data Analysis
MPHP	Ministry of Public Management, Home Affairs, Posts and Telecommunication
MPM	Market Price Method
NACSJ	Nature Conservation Society of Japan
NFFCA	National Federation of Fisheries Cooperative Associations
NOAA	National Oceanic and Atmospheric Administration
NPV	Net Present Value
NSB	Net Social Benefits
NSKF	Nagasaki, Saga, Kitakyushu and Fukuoka

NTT	Nippon Telegraph and Telephone West Corporation
OE	Open-ended
PVB	Present Value of Benefits
PVC	Present Value of Costs
RST	Remote Sensing Techniques
TCM	Travel Cost Method
TDL	Tokyo Disney Land
TDS	Tokyo Disney Sea
USJ	Universal Studios Japan
VIF	Variance Inflationary Factor
WTP	Willingness to Pay

PART I
OVERVIEW

CHAPTER 1 Introduction

1.1 Measuring Environmental Benefits and Costs

We want our environment to be safer and cleaner. But how clean or safe it should be? How clean we should make the sea or air? Should we build a new dam for flood protection? Or we just tolerate the flood to get rid of the negative impact of the dam on the eco-system? Like these questions- nothing comes for free. If we want to get something, then we got to forgo something. Same with environmental resources, which are entirely public goods and are limited. So before making any decision regarding the use of such public goods, we need to compare the benefits of defending environmental resources compared with the opportunity costs or benefits forgone for alternative uses, i.e., conducting cost-benefit analysis (CBA). Accordingly, economists rely on empirical research in the form of benefit-cost analysis to aid decision makers for arriving at more informative and economically efficient choices by balancing the costs of environmental goods against their benefits (Mitchell and Carson 1989).

Efficiency is attained when the *marginal benefit* equals the *marginal cost* (Tietenberg 1996). This simple form of maximum economic efficiency which can be achieved by cost-benefit analysis is explained by Fig. 1.1. In the figure, the total benefit increases as the quantity (Q) increases but at a decreasing rate. In contrast, total costs increases at an increasing rate. The net benefit is maximized when Q* is supplied, where the marginal benefit is equal to marginal cost, so that total benefits surpass total costs by the largest possible amount. Regarding efficiency, moving from Q_0 towards Q* increases efficiency; i.e.,

$$NPV(Q^*) > NPV(Q_1) > NPV(Q_0)$$

Whereas, moving beyond Q* reduces efficiency and the net benefit becomes zero at allocation Q_2. At this point the average benefits equal average costs, total benefits equal total costs, and the benefit-cost ratio equals to 1. By using benefit-cost ratio or positive net present-value (NPV)[1] decision rule can lead towards selection of inefficient choices. Only the maximum net present-value approach is completely attuned with efficiency.

[1] See Appendix B for the present value, net present value, i.e., time value of money concept.

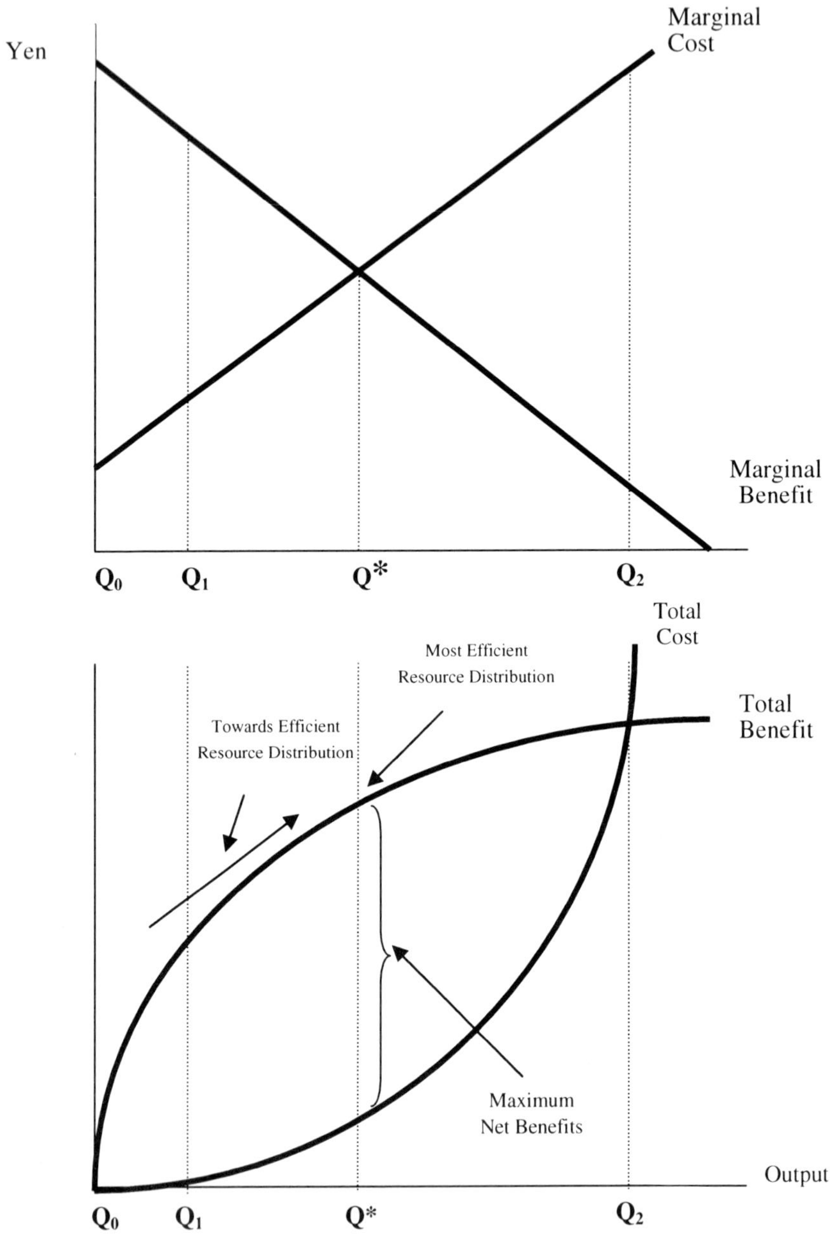

Fig.1.1. CBA aims towards more efficient distribution of resources

Thus maximization of net-present value by equaling marginal benefits with marginal cost through benefit-cost analysis can ensure most efficient distribution of

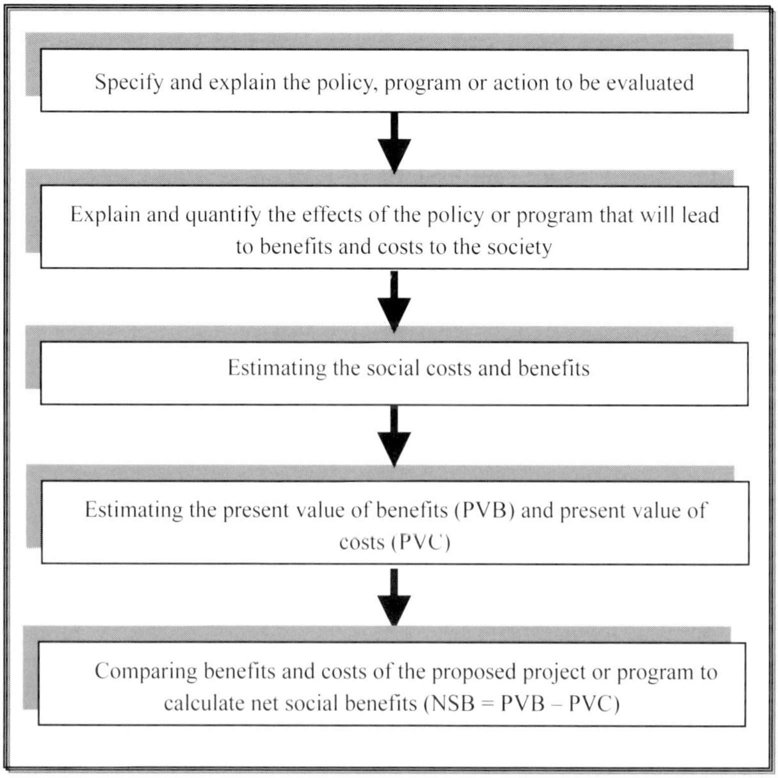

Fig. 1.2. Steps involved in cost-benefit analysis

our scarce environmental resources. But, this requires enumerating and evaluating all of the measurable benefits and costs and comparing them which leads us to our next problem, that is, how we can measure environmental benefits and costs? This difficulty arises as environmental goods, e.g., clean air or wetlands are not sold in the common market place. Thus for valuing non-marketed environmental values more efforts are required due to the absence of well-defined markets like other goods and services, where prices are revealed openly (Lesser et al. 1997). We need individuals or markets to disclose environmental values to us. The focus of this book is to use and develop methods to create individuals or markets so that we can measure environmental benefits and costs, and finally can conduct cost-benefit analysis to make more informative decision regarding the use and conservation of environmental resources.

1.2 Basic Steps in Cost-benefit Analysis

In order to ensure efficient allocation of scarce environmental resources, cost-benefit analysis should be conducted for each policy, program or action involving such resources. Conducting cost-benefit analysis of environmental amenities involves the steps briefly described in Fig. 1.2. It involves the specification of the policy or program under consideration and quantification of all the benefits and costs expected to arise from the initiation of the project including environmental benefits and costs. These will lead towards the derivation of present value of social benefit (PVB) and present value of costs (PVC) of the project to calculate the net social benefits (NSB).The choice resulting from such analysis might not be the most socially acceptable option, but it is one of the best options available for inserting public preferences into decisions involving public goods.

1.3 Types of Environmental Values

The environmental values that are to be estimated for inserting into CBA are classified into two broad categories: use and non-use (also referred to as 'passive use') values. As the name implies, use value is defined as the value that individuals place on environmental goods and services they actually consume. For example, bird watching, hunting, fishing or visiting some recreational spot. On the other hand, non-use value is defined as the value that is not associated with the actual consumption of the environmental goods and services. For example, a person might be willing to pay for protecting the Isahaya Bay Wetland, even though he or she never been or even wants to go there, but simply because they want it to be there. Use and non-use values can further be subdivided into different categories, as shown in Fig. 1.3.

1.4 Environmental Valuation Techniques

There are two major types of environmental valuation techniques: monetary and non-monetary. As the name implies, monetary valuation techniques (sometimes called direct valuation techniques) value environmental goods and services on the basis of the monetary values that individuals place on receiving or avoiding them. But it does not necessary require that the good or service must be sold in the market. Rather it only requires estimating how much purchasing power people would be willing to give up to get it (or would need to be paid to give it up), if they were forced to make a choice.

Conversely, non-monetary valuation techniques (also called indirect valuation techniques) values environmental resources without placing a direct monetary value on them. These methods are particularly useful when individuals are unaware of the environmental impact in question and to make decisions based on ranking or prioritizing the expected benefits of environmental investments (Lesser et al. 1997). A brief description of the major monetary and non-monetary valuation techniques are provided in Table 1.1.

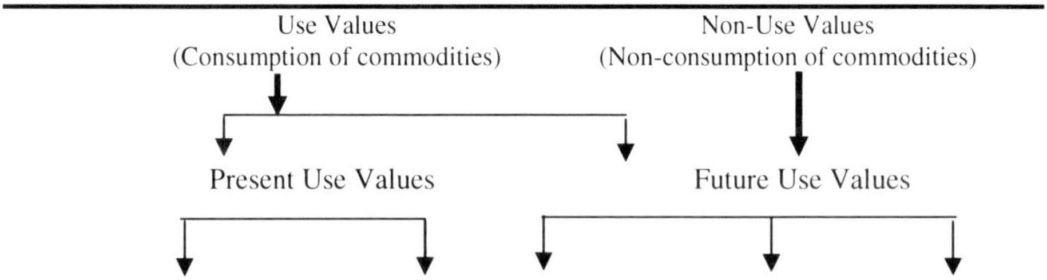

Name	Direct use Values	Indirect use Values	Optional use Values	Bequest Values	Existence Values
Description	Value is generated from direct consumption of commodities.	Value is generated from indirect consumption of commodities.	Value is generated from the options to consume commodities directly or indirectly in future.	Value is generated from conserving of consumption opportunities for future generations.	Value is generated from preserving existence of biological resources from ethical viewpoint.
Examples	Food, fiber, timber, recreation, etc.	Maintenance of ecological systems, nutrient recycling, etc.	Use of biodiversity in future pharmaceutical products, use of genes in increasing future agricultural productivity, etc.	Preserving unique habitats, species, etc.	Preserving endangered habitats, species, etc.

Fig. 1.3. Types of environmental values

Source: Compiled and adapted from Lesser et al. (1997)

1.5 A Comparison of Methodologies

All the methodologies discussed have their relative advantages and disadvantages. There is no perfect environmental valuation technique. One technique might be appropriate for one situation and inappropriate in others. However, based on the existing literature CVM appears to be reasonable, though potentially biased, methodology (Lesser et al. 1997). Careful designing and execution can minimize various biases resulting from CVM, which has the rare advantage to estimate both use and non-use values. Hence CVM studies conducted with enough care can provide reliable estimates. Accordingly, this particular method is elaborated in details in the following chapters and used in case studies as the environmental valuation technique.

Table 1.1. Environmental valuation techniques

Valuation Technique	Ways to value environment	Applications	Limitations
A. Monetary Valuation Techniques:			
1. Market Price Method (MPM)	Estimates economic values from the economic benefits and costs arising from changes in environmental quality.	Can be applied to estimate the economic value of ecosystem products or services that are bought and sold in commercial markets	Market data may only be available for a limited number of goods and services provided by an ecological resource and may not reflect the value of all productive uses of a resource.
2. Hedonic Pricing Method (HPM)	Estimates economic values for ecosystem or environmental services that directly affect market prices of some other goods. Most commonly applied to variations in housing prices that reflect the value of local environmental attributes.	Most commonly applied to variations in housing prices that reflect the value of local environmental attributes.	The scope of environmental benefits that can be measured is limited to things that are related to housing prices.
3. Travel Cost Method (TCM)	Estimates economic values by assuming that the value of an eco-system or recreation site is reflected in how much people are willing to pay to travel to visit the site.	Often applied to value recreation sites, which have zero or nominal admission price	Can not estimate non-use values
3. Contingent Valuation Method (CVM)	Estimates economic values by asking people to directly state their willingness to pay for specific environmental services, based on a hypothetical scenario.	Can be applied to estimate economic values for all kinds of ecosystem and environmental services and most widely used method for estimating non-use values	Prone to provide biased result and attempts to reduce this makes the study expensive and time consuming
4. Contingent Choice Method (CCM)	Estimates economic values by asking people to make tradeoffs among sets of ecosystem or environmental services or characteristics. Does not directly ask for willingness to pay—this is inferred from tradeoffs that includes cost as an attribute.	Can be used to estimate economic values for virtually any ecosystem or environmental service, and can be used to estimate non-use as well as use values	Due to the unfamiliarity of some tradeoffs, respondents may find it difficult to evaluate.
B. Non-monetary Valuation Techniques:			
Estimating Dose-Response Function	Uses dose and response to indirectly measure the effect of certain environmental changes	Can be applied to measure the effects of air pollution on human health, agricultural production, or corrosion of buildings and can form the basis for monetary valuation	Degree of precision of measurement is questionable.

Source: Compiled and adapted from Lesser et al. (1997)

CHAPTER 2 Contingent Valuation Method

2.1 Introduction

In simple terms, contingent valuation is a method of estimating the value that people place on a particular good. It uses survey techniques to elicit people's willingness to pay (WTP) to obtain a particular good or willingness to accept (WTA) to give away the good[1]. It can be applied for goods both which are and are not traded in regular marketplaces. For goods, which are not traded in the marketplace, like restoring the Isahaya Bay wetland, a hypothetical marketplace is created in which respondents are given the opportunity to buy the good. Since the elicited WTP values are contingent upon the particular hypothetical market described to the respondents, this approach came to be called the contingent valuation method (Brookshire and Eubanks 1978). At different times and in various places, the contingent valuation method is also called the survey method, the interview method, the direct interview method, the direct questioning method, the hypothetical demand curve estimation method, the difference mapping method, and the preference elicitation method (Mitchell and Carson 1989).

2.2 The Development of CVM

There is controversy regarding the first use of contingent valuation method (CVM). According to Mitchell and Carson (1989), " the contingent valuation method first came in to use in the early 1960s when economist Robert K. Davis used questionnaires to estimate the benefits of outdoor recreation in a Maine back woods area of USA". Whereas, Hanemann (1992) traces the empirical roots of CV to 1958 and a US National Park Service- funded study of outdoor recreation in the Delaware River Basin area, USA. However, whatever be the exact year or study, it is clear that CVM first came into use in the 1960s. The method was used only infrequently during the 1960s and exclusively in the US. Later in 1970s CVM saw a gradual sustained growth and by the end of the decade got official recognition by the US Water Resources Council as a recommended valuation technique (Bateman

[1] The WTA format should usually be avoided in CV studies because it does not elicit valid data under many circumstances (Mitchell and Carson 1993).

et al. 1999). In this decade this method was also applied to the European countries (Bohm 1972).

A huge number of studies based on CVM were conducted in 1980s. The first application of it in the developing countries took place in 1985 (Whittington et al. 1990). The decade also witnessed the incorporation of CVM in the institutional decision making under the Comprehensive Environmental Response, Compensation and Liability Act (CERCLA) 1980 in the USA, as a technique to measure the extent of natural resource damages from the spillage of hazardous substances with reference to the Exxon Valdez incident. This study is particularly important because of its high profile, careful design, implementation and naturally attracted lot of attentions (see Box. 2.1 and Figs. 2.1through 2.3 for details of the Exxon Valdez study).

The CVM enters into 1990s with higher academic concern and institutional approval in different parts of the world. Accordingly, National Oceanic and Atmospheric Administration (NOAA) formed a 'Blue-Ribbon Panel' of independent economic experts (including Nobel laureates Kenneth Arrow and Robert Solow) to evaluate the use of CVM for determining passive use values (Titenberg 1996). The report of the experts, issued on January 15, 1993 (58 FR 4602) cleared the method as sound only if studies were conducted by following the set guidelines which are explicitly spelt out. The panel concluded that:

" can produce estimates reliable enough to be the starting point of a judicial process of damage assessment, including lost passive values.....[A well-constructed contingent valuation study] contains information that judges and juries will wish to use , in combination with other estimates, including the testimony of expert witnesses."
[Source: Adapted from Titenberg 1996]

However, the bibliography prepared by Carson et al. (1995) of CVM studies contains over 2,000 studies, but few of these satisfies the guidelines set by the NOAA panel. Finally, we are now in the year 2005, with the sustained rise in academic interest and institutional acceptance expanding the use of CVM in almost all over the world. But this does not indicate that the debates on CVM have ended, still continuing and lot to be witnessed in the years to come.

2.3 Survey Techniques

The contingent valuation method (CVM) relies on survey techniques to value the particular commodity of interest. Various survey methods are available such as in-person interviews, telephone survey, mail survey etc. They all have their relative advantages and disadvantages. There is no single best method, which provides better results for all types of questions. Some most commonly used survey techniques are summarized in Table 2.1.

On the night of 24 March 1989, the oil tanker Exxon Valdez ran aground in Prince William Sound in south-eastern Alaska, releasing some 11 million gallons of crude oil. Over the next six months this, the largest oil spill in US waters to date, resulted in the death of 36,000 sea birds, 1,000 sea otters, and over 150 bald eagles (Maki 1991). This incident set off a major debate as to the validity of CVM in determining the monetary liability of the Exxon Company associated with the spill.

Following Comprehensive Environmental Response, Compensation and Liability Act (CERCLA) guidelines, the state of Alaska gathered some renowned researchers to undertake an extensive and through CV assessment of the non-use damages caused by this disaster. Concerned of the size of the damage resulting from such study, Exxon Company also employed a number of equally eminent economists to investigate the validity of the CV technique.

As a result, the State of Alaska valued the lost passive values to some $2.8 billion (based on median WTP values of $31 per household; Carson et al. 1992). Accordingly the Exxon Company reacted by seeing the prospect of paying a huge amount as non-use compensation and presented their side of the explanation by showing critical comments regarding CV studies to estimate non-use values (Cambidge Economics 1992; Hausman 1993).

All these debates and arguments influenced the final out-of-court settlement negotiated before the completion of the various CVM studies and Exxon agreed to pay $1.1 billion in damages, on top of the $2.5 billion cleanup costs associated with the spill. Although no final decision as to the accuracy of CVM for calculating passive-use values arose from this Exxon Valdez case, it brought CVM to the legal and economic scrutiny and center of various debates.

Source: Reproduced and compiled from Bateman et al. 1999 and Lesser et al. 1997.

Box 2.1. CVM and the case of the Exxon Valdez

While in-person interviews are obviously the technique of choice for contingent valuation surveys, experience with telephone and mail CV surveys suggest that, with the exception of the sample non-response bias problem in mail surveys, their shortcomings may be overcome if the respondents are familiar with the amenity, or if the scenario is relatively simple (Mitchell and Carson 1993). In the study of the Isahaya Bay wetland, mail survey technique was used, because respondents were reasonably familiar about the good to be valued through mass media, as evidenced from the questionnaire pre-testing.

Fig. 2.1. *Exxon Valdez* tanker surrounded by booms near Naked Island.
EVOS photo by Pamela Bergman

Fig. 2.2. Cleanup workers spray oiled rocks with high pressure hoses.
EVOS photo by Roy Corral

Fig. 2.3. An oiled seabird struggles on the beach.
EVOS photo

Table 2.1. Comparison of the Survey Techniques

Subject	In-Person Interview	Telephone Survey	Mail Survey	Internet Survey
Cost	Very high Depends on questionnaire length and geographical spread	High Depends on questionnaire length and call backs	Low Depends on number of follow-ups	Low Marginal costs very small
Ease of Access to Respondents	Medium Depends on availability of lists and access	Very high Random digit dialing	High Depends on availability of appropriate lists	Low "Spamming restrictions require panels of willing respondents
Interviewer Bias	High Personal presence, monitoring difficult	Medium Interviewer cues	Low Uniform presentation	Low Uniform presentation
Availability of Information	Very high Interactive communication and visual aids possible	Low Verbal communication limits complexity of content	High Visual aids possible	Very high Visual aids and interactive questions possible
Application to Developing Countries	Very high Relatively easily accessible to respondents	Low Limits the number of respondents	High Depends on availability of appropriate lists	Low Limits the number of accessible respondents

Source: Reproduced from Boardman et al. 2001

2.4 Elicitation Methods

There are different types of elicitation methods for deriving the willingness to pay (WTP) value. Among them especially dichotomous-choice (DC) format is the most widely used one. Follow-up questions are also used to increase the precision of the estimate with DC question. The NOAA blue ribbon panel also advocated this method as the most appropriate one in most circumstances (Arrow et al. 1993). While among others, bidding game (BG) approach has been almost deserted because it tends to result in a starting point bias. A summary of the characteristics of the most commonly used elicitation methods is provided in Table 2.2.

Table 2.2. Commonly used Elicitation methods in CV survey

Subject	Open-ended/ Direct question	Bidding Game	Dichotomous-choice method	Double Bounded Dichotomous-choice method
How-To's	Here the respondents are asked to state their maximum WTP for the amenity to be valued	Here the respondents are asked whether they would pay a specified amount for the amenity to be valued and the amount is lowered or raised like auction depending on the negative and affirmative answer.	Here the respondents are asked if they are willing to pay single randomly assigned amount for the amenity on all-or-nothing basis.	In addition to the question in the DC method, here the respondents are asked an additional question if they would pay a higher (in case the first answer was YES) or lower (in case the previous answer was NO) amount.
Number of Question	Single	Iterated series of question	single	Iterated series of question
WTP Obtained	Actual	Actual	Discrete indicator of WTP	Discrete indicator of WTP
Major Advantage	Provides Straightforward actual valuation of amenities	Can draw out the actual WTP with guided series of questions	Simplify the respondent's valuation choice as questions are easy to answer	Offers the potential for considerable gains in efficiency
Major Disadvantage	Might provide unrealistic responses	Prone to starting point bias	Requires relatively larger sample size to achieve accurate results	Danger that the respondents exposure to the first offer would influence them to accept the follow-up offer

2.5 Biases

The major constraint in the use of the contingent valuation method is that generally there is possibility that the respondents will give biased answers (Tietenberg 1996). The following are the possible sources of bias in CVM studies:

1. *Strategic Bias*: Occurs when respondent provides or the scenario of the questionnaire provokes him or her to provide a biased answer to influence a particular outcome.
2. *Information Bias*: Arises when respondents are forced to value certain environmental goods or services of which they have little or no experience.

3. *Starting Point Bias*: Arises when survey instrument provides a predetermined range of choices for answering their values.
4. *Hypothetical Bias*: Arises when the respondents do not take the study seriously by realizing that he or she will not have to pay actually.
5. *Sampling Bias*: Arises due to improper sampling design and execution.
6. *Non-response Bias*: Occur due to lower response rate.

2.6 Criticisms

The CVM is not also beyond any criticism. It has the shortcomings that survey studies usually have. The way the WTP questions are made may biases value estimates. Also respondent's reply to survey questions about their behavior often differs from what they actually do. Much controversy surrounds the use of CVM when most of the value of the good derives from passive use, as has been typical in litigation over the damages to natural resources and amenities caused by releases of pollutants (FAO 2000). CVM values derived are contingent upon the level of information provided by the survey and information supplied by the respondent (Pate and Loomis 1997). In addition, moral and ethical positions are also found to be important in CVM studies, as people's decision to engage in the monetary valuation process is associated with her/his fundamental ethical beliefs (Spash 2000).

Despite of these criticisms CVM has been widely used by researchers both in developed and least developed countries. The following chapters of this book apply this technique to conduct cost-benefit analysis of various selected case studies.

PART II
CASE STUDY 1: THE ISAHAYA BAY WETLAND (IBW)

CHAPTER 3 Overview of the IBW

3.1 Introduction

Wetlands, such as mires and tidal flats, form a unique ecosystem, providing the habitat for various animals and plant species and of immense economic importance. There are 37 tidal flat wetlands in Japan, comprising of 335 km^2 in 1998 (reduced from 510 km^2 in 1991). Currently, these wetland areas are sharply reducing as development projects are undergoing on 16 of these wetlands (NACSJ 1998). One of such important and endangered wetlands is the Isahaya Bay Wetland (see Fig. 3.1 for the location). This wetland is located in Isahaya Bay, a part of the Ariake Sea - quiet pocket bay in Nagasaki Prefecture, Western Japan, approximately extending to 100 square km. The Ariake Sea is famous as the home of mudskipper. It is also the biggest center of seaweed production[1], providing about two-fifths of the nori production of the nation (Mainichi 2001a). Also a total of 144 species of water bird including 60 species of shorebird have been recorded.

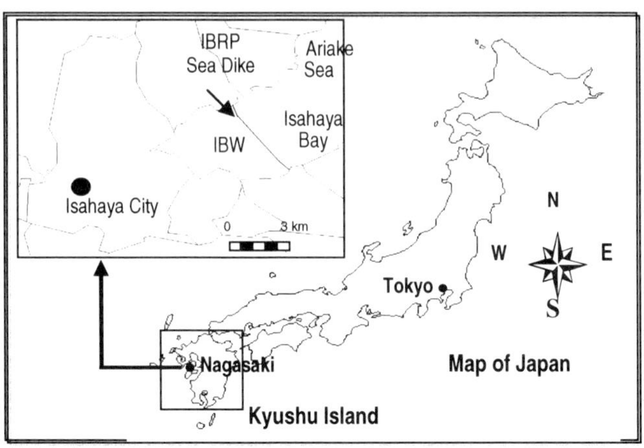

Fig. 3.1. Location of the IBW and IBRP

[1] Greenish-black processed seaweed, called *Nori* in Japanese, is a staple of Japanese dishes - wrapping specialties from rice balls to sushi.

Before the start of the IBRP (April 15, 1988)

During the project work of the IBRP (April 14, 1999)

Fig. 3.2. Satellite images of the IBW and IBRP

Fig. 3.3. Sea dikes of the IBRP
Photo by F. Iguchi

IBW is the largest tidal flat wetland in Japan, comprising about 6% of the total country's wetland. Its depth is in 10-meter range, and tidal differences is about 6 meters. Its 3,000 ha (hectares) of muddy tidal flats are composed of fine silt with a high proportion of organic matter and are considered as one of the country's richest marine area in terms of biological productivity. However, recently the Isahaya Bay Wetland became a controversial issue due to the construction of sea dikes on it and closing off 3,500 ha of the bay by a 7 km long seawall (Asiaweek 2001). Figures 3.2 and 3.3 explain the before and after project impact of the IBRP on the IBW. The objectives of this $1.8 billion project were land reclamation and flood prevention.

The history of land reclamation in this region dates back to ancient time. The extensive plains landward of the upper part of Isahaya Bay were formed over the

Table 3.1. Chronological events of the Isahaya Bay Reclamation Project

Date	Events	Remarks
1952	"Nagasaki Large-scale Land Reclamation Project" was drafted as a part of the "Ariake Sea Comprehensive Development Plan," to close off the entire Ariake Sea in order to develop more land for farming rice.	Eventually, both plans were terminated due to abundance of rice and as well as financial difficulties.
1982	The above project was revived by adding flood prevention as one of the objectives with land reclamation and renamed as, "Isahaya Bay Reclamation Project (IBRP)".	The area to be reclaimed was reduced from about 10,000 ha to 3,550 ha, due to strong protests from the fishermen.
1986	First original draft plan was finalized for the IBRP with probable Cost-benefit Analysis (CBA) and started the construction work.	Reclaimed Area: 3,550ha (New Farmland: 1,840 ha, Catch Basin: 1,710 ha). Expected Year of Completion: 2000.
1996	Modification in project plan due to increase in costs (44 percent and 56 percent for increase in raw-material cost and change in construction plan, respectively).	Expected Year of Completion was revised to year 2006.
1997	A 7,050-meter (7 km.) long dyke across the center of the Bay was completed and dropped 293 iron slabs cutting of the IBW from the sea.	Right from this moment attracted a lot of criticisms and protests from local fishermen, environmentalist and so on.
2001	The government agreed to suspend the reclamation work and investigate by an expert panel regarding the impact of IBRP on the eco-system of the Ariake Sea (March 2001).	The panel also recommended that the water gates should be opened to investigate the cause for poor seaweed in recent years.
2001	Protest from local fishermen, environmentalists and opposition politicians forced the MAFF to re-modify and scale down the project (August 2001).	Scaled down the reclaimed area to 700 ha which have been almost completed by this time.
2002	Seaweed growers blocked workers from continuing the reclamation of the Isahaya Bay (January 2002).	They blamed the government for continuing the reclamation work in spite of expert advice to stop it.
2003	Future of the project and restoration of the Isahaya Bay Wetland are still undecided.	Construction work in the project site is still going on.

Compiled from:
1. Japan Wetland Action Network (JWAN), National Report, May 1999.
2. Miyairi 1998.
3. Various issues of Mainichi Daily News 2001~2002.
4. Various Issues of Asahi Shimbun 2001.

The project has two objectives: (1) produce highly productive farmland; (2) prevent disaster in the area.

(1) Reclamation of highly productive farmland

One of the objectives of the project is to promote agriculture by reclaiming large areas of highly productive farmland in Nagasaki Prefecture, which is mountainous and suffers from the shortage of farmland of good quality. Farmers growing vegetables in the area will be able to increase their average farming size from 1.0 ha to 3.0 ha by obtaining 2.0 ha of land in the reclaimed area, enabling them earn the target income estimated in the "New Agricultural Policy of Nagasaki Prefecture, 1996".

(2) Disaster Prevention

The other objective of the project is to prevent disaster in the Isahaya Bay area, which has suffered from several disasters caused by high tide, flood and poor drainage. The construction of the sea dike and reservoir for flood control will prevent such disasters as follows:

(a) The sea dike completed under the project could prevent the area from being damaged even by a high tide in one of the largest typhoons (Ise Bay Typhoon in 1959).

(b) Flood water from the rivers would flow into the regulating reservoir even if both Ise Bay typhoon and Isahaya Flood (one of the biggest disasters caused by flooding (1957)) take place simultaneously, because the sea dike will prevent the water level of the reservoir from rising due to a high tide.

(c) Drainage in the lower areas can be easily done by channeling excess water under normal rains into the reservoir without being influenced by tides.

Source: Message from the Agricultural Structure Improvement Bureau, Japanese Ministry of Agriculture, Forestry and Fisheries.

URL: www.maff.go.jp/soshiki/koukai/sekkei/isahaya-e.html

Box 3.1. Objectives of the Isahaya Bay Reclamation Project

course of 2000 years through both natural and artificial land reclamation; the latter began about 600 years ago. The methods used to reclaim the land during the Edo era (1603-1867) were simple, gradual and effective, and followed the dictates of nature (JWAN 1999). Chronological list of developments of the Isahaya Bay Land Reclamation project on the Isahaya Bay Wetland are listed in Table 3.1.

Table 3.2. Comparative cost – benefit analysis of the project

	First draft plan of 1986	Changed plan of 1996	Revised Plan of 2001
Reclaimed Area	3,550 ha (New Farmland: 1,840 ha, Catch Basin: 1,710 ha)	3,550ha (New Farmland: 1,840 ha, Catch Basin: 1,710 ha)	700 ha
Project Cost	¥135 billion	¥237 billion	¥246 billion
Expected Year of Completion	2000	2006	2006
Benefit-Cost (B/C) ratio	1.03	0.58	0.83

3.2 Objectives of the IBRP

According to the Agricultural Structure Improvement Bureau, Japanese Ministry of Agriculture, Forestry and Fisheries, the objectives of the Isahaya Bay Land Reclamation Project were as shown in the Box 3.1.

3.3 Cost – Benefit Analysis of the IBRP

In 1989, the project was estimated to cost ¥135 billion; this had escalated to a staggering ¥246 billion. It is expected to be completed in 2006 with a cost of over $2 billion for creating 1,500 ha of new farmland (Asiaweek 2001). That means $1.3 million for each ha. A summary of the frequently modified cost-benefit analysis of the IBRP is provided in Table 3.2.

3.4 Current Status and Controversy

The IBRP was widely criticized when the gates were closed initially in 1997, threatening the ecosystem of the bay, which was formerly called the "womb of the Ariake Sea.". Local fishermen blame the reclamation project for serious ecological disaster in the region. Seaweed farmers alleged that since a set of dikes sealed off part of the bay for reclamation in 1997, the sea has been polluted by an inflow of nutrient-rich waters, resulting in a poor seaweed crop (Mainichi Shimbun 2001a). The farmers are demanding the government to cancel the reclamation project and open the water gates along the dikes to allow the environmental condition of the

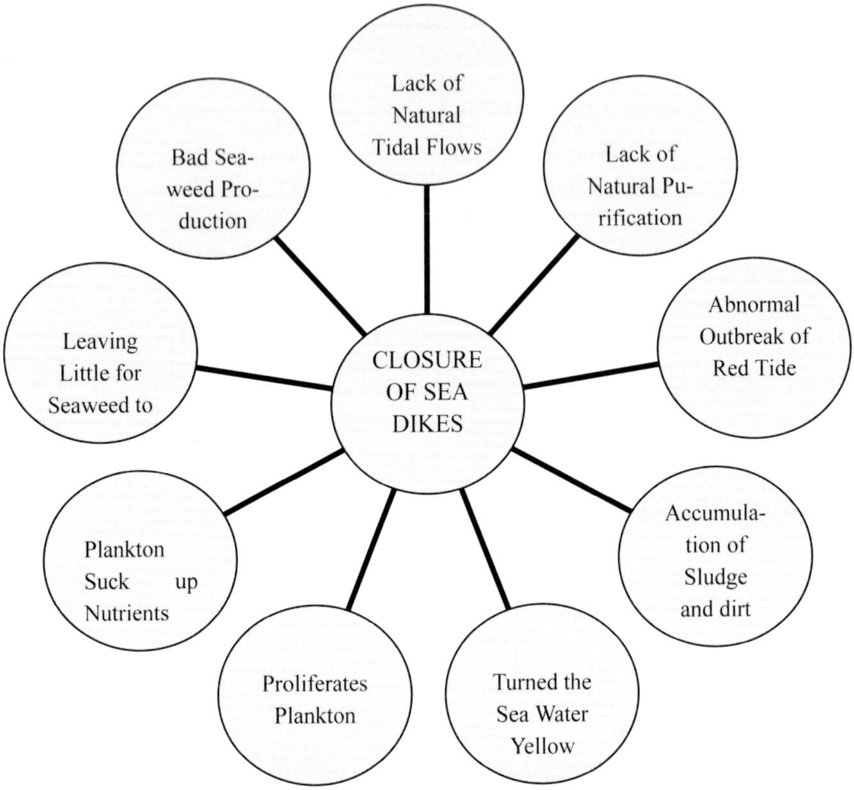

Fig. 3.4. Effect of closing of sea dikes

area to return to normal- where the water is purified naturally by the tideland ecosystem (Mainichi Shimbun 2001a).

Also the environmentalists and opposition politicians dubbed the wall a "guillotine that killed nature" and say it led to pollution in the sea outside the gates, caused harmful red tides and thus the huge drop in seaweed production. Four years later, the fertile tideland is dried up. The process by which the seaweed farmers and environmentalists argue that the seaweed production went bad can be explained by the Fig. 3.4.

Whatever should be the cause, production of seaweed in the first 10 months of fiscal year 2000 totaling about 3.8 billion sheets, a 25 percent drop from the same period a year earlier, according to the National Federation of Fisheries Cooperative Associations (NFFCA) (Mainichi Shimbun 2001b). Again, output in early January 2001 was only 77 percent of what it was a year earlier in Kumamoto, about 70 percent in Saga and Nagasaki, and slightly above 50 percent in Fukuoka. At a recent auction in Kumamoto, the low-quality crop that had attracted no buyers the last time was put on sale again because there was no fresh harvest.

However, Agriculture, Forestry and Fisheries Ministry (MAFF) contend that there is no direct correlation between the project and the poor crops (Asahi Shimbun 2001). They identified fall in average yearly rainfall and rise in average temperature as the reasons for poor seaweed production (Mainichi Shimbun 2002). Due to the growing protests, however, a survey was launched in March 2001, to assess the current condition and is expected to start a fresh study to judge the impact of the project (Mainichi Shimbun, 2001c). Finally, the agriculture ministry commented that the IBRP would be reassessed again, but at the same time maintained that no link exists between reported environmental problems and the land reclamation. The reclamation of what was Japan's largest tidal wetland is scheduled to be completed by 2006 (Asiaweek 2001). By that time, however, the area may have become a "dead sea", as the national daily Asahi Shimbun mentioned in its editorial.

3.5 Important Questions

In the wake of the controversies between local fishermen, environmentalists, and government, it is of great interest to seek answer to the following questions:

1. How much the Isahaya Bay wetland is worth to the people living in different parts of Kyushu Island, in monetary terms?
2. What types of cost-benefit analysis (CBA) methods were undertaken to justify the initiation of IBRP?
3. How socio-environmental aspects to be affected by IBRP were considered in the analysis?
4. Is there any way to consider socio-environmental aspects before initiating a project to avoid future conflicts?

In order to answer the above questions, a survey study was conducted. The following chapters demonstrate the findings of this study.

CHAPTER 4 Environmental Valuation of the IBW

4.1 Introduction

As we have observed at the end of Chapter 3 that, in order to protect or restore the Isahaya Bay wetland (IBW) it is important to value its worth, this chapter aims to estimate the environmental value of the Isahaya Bay Wetland using contingent valuation method (CVM) and to compare the total WTP with the variation in WTP calculated on the basis of closeness to the project site, age and income of the respondents.

It should be mentioned here that, the objective is to estimate the value of the Isahaya Bay Wetland and not the value of the Isahaya Bay reclamation project. Enough care was taken in designing the survey instruments so that the respondents of the study do not confuse these two interrelated terms.

4.2 Valuation Methodology

Contingent Valuation Method (CVM) is used to estimate the willingness to pay (WTP) value, which depends on survey techniques to value a particular commodity. In order to choose from the various survey techniques and elicitation methods, particularly suitable for our study of valuing the Isahaya Bay Wetland, we have conducted pre-testing of questionnaires. And this led us to opt for mail survey technique and dichotomous choice (DC) elicitation method for the final questionnaire survey. Both parametric and non-parametric estimate analysis were conducted.

In this study double bounded dichotomous-choice (DC) elicitation method were used for deriving the WTP figures. Again under DC method both parametric and non-parametric estimate analyses were conducted to estimate WTP. We consider the non-parametric estimation of the distribution function F of a real-valued random variable *X*, when the sample data are incomplete due to restricted observation brought about by grouping, censoring and/or truncation (Turnbull 1976). Among non-parametric estimation Turnbull method is followed, which uses the following expression formula:

$$LL = \sum_{i \in yy} \ln S(T_{hi}) + \sum_{i \in nn} \ln[1 - S(T_{li})] + \sum_{i \in ynorny} \ln[S(T_{li}) - S(T_{hi})] \quad (4.1)$$

Where,
LL= the maximum likelihood estimate
$S(T)$ = the probability to accept bid value T
T_{hi} = highest bid value
T_{li} = the lowest bid value to the *ith* individual
yy = the set of respondents who answered *yes* for both the bid values
nn = the set of respondents who responded both time *no*
yn = the set of respondents first time *yes* and then *no*
ny = the set of respondents first time *no* and then *yes*

Whereas, under Weibull method of parametric estimation, the value of μ and σ are fixed to give shape to the cumulative distribution function and determines $S(T)$ by the following equation:

$$S(T) = \exp\left[-\exp\left(\frac{\ln T - \mu}{\sigma}\right)\right] \quad (4.2)$$

Where, exp is the exponential operator, μ is the mean and σ is the standard deviation. After determining the $S(T)$, then we proceed for the estimation of LL.

Both of these methods have their relative advantages and disadvantages. Hence, we are going to conduct our calculation by using them together and compare the results.

4.3 Justification for Using CVM

Now the question arises as to why we choose to use contingent valuation method (CVM) among all other valuation techniques available for quantifying the value of the Isahaya Bay Wetland (IBW)? The reason is that the CVM can be used to estimate economic values for almost all kinds of ecosystem and environmental amenities, and can estimate both use[1] and non-use[2] (passive) values. It is considered both as an alternative method of valuation to travel-cost (TC) and hedonic pricing (HP) models and as being able to quantify some types of benefits, such as non-use or passive-use benefits, which lie outside the scope of TC and HP studies (Bateman et al. 1999). Accordingly, our case of this study (IBW) being a non-marketed

[1] Value derived from actual use of a good or service. Uses may include indirect uses.
[2] Value that are not associated with actual use, or even the option to use a good or service.

environmental amenity having both use and non-use values are expected to be effectively valued by CVM on the basis of the set guidelines.

Many researchers also have used CVM for valuing wetlands. They have suggested transforming the value of the wetlands and other environmental entities into monetary amounts for considering interactions with the environment as the most appropriate way (Spash 2000). According to Costanza et al. (1989) 'the estimation of the economic value of natural wetlands is a difficult and complex task, but it is essential to rational management', and they went to do so by applying contingent valuation method (CVM) as one of the methods. Among many others, Loomis et al. (1991), Pate and Loomis (1997) and Kuriyama (1998), also applied CVM for valuing wetlands.

4.4 Pre-testing Study

4.4.1 Introduction

Before proceeding into the main study, in this section of the study, we tried to develop the base for designing appropriate questionnaire by selecting the more accurate Contingent Valuation (CV) elicitation method. Accordingly, three pre-testing trial results were compared, which used different elicitation methods and alternate order of the questions in the questionnaire.

4.4.2 Summary of Pre-testing Questionnaire

The questionnaire used in the study contained three parts (see Fig.4.1). The pretesting was conducted under different elicitation methods and different order of questions in the questionnaire to compare the variation of results. For this purpose, we used two-elicitation methods viz., bidding game (BG) in the first trial and dichotomous choice (DC) payment method in the second and third trial, with follow-up questions. As shown in Fig.4.2, in the first and the second trial, we asked the reasons for not willing to pay before going to the bidding values. But this sequence of questioning was changed in third trial. This has provided us an interesting result regarding the answering attitude of the respondents.

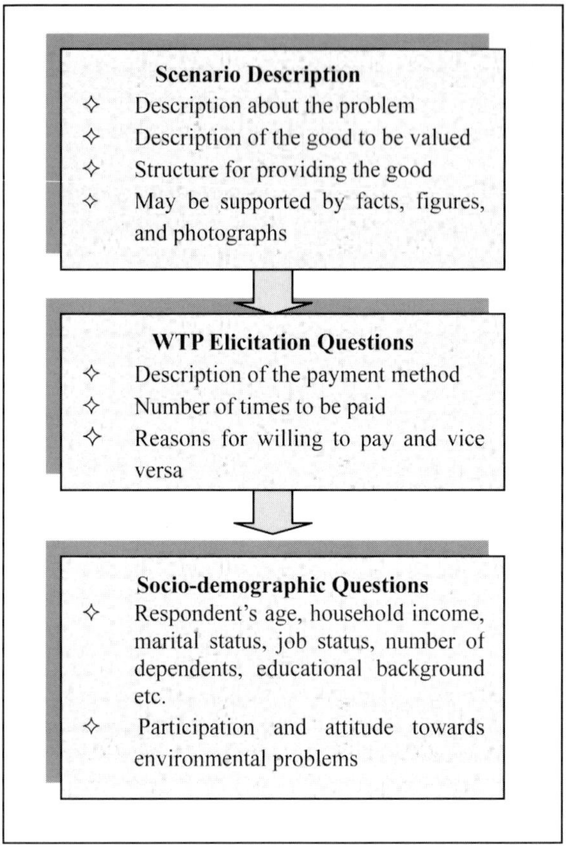

Fig.4.1. Structure of the questionnaire

4.4.3 Comparisons of Pre-testing Results

In the first trial, a total of 17 respondents were questioned, in the second trial, a total of 42 respondents and 104 respondents in the third trial (see Table 4.1). Only one edition of questionnaire was used in the first trial, with a flat rate of ¥5000 and bidding game elicitation method was used. While in the second and third trial, three editions of the questionnaire were used, with a starting amount of ¥1000, ¥5000 and ¥10,000 respectively. The results are summarized through Table 4.1 and Figs. 4.3 ~ 4.4.

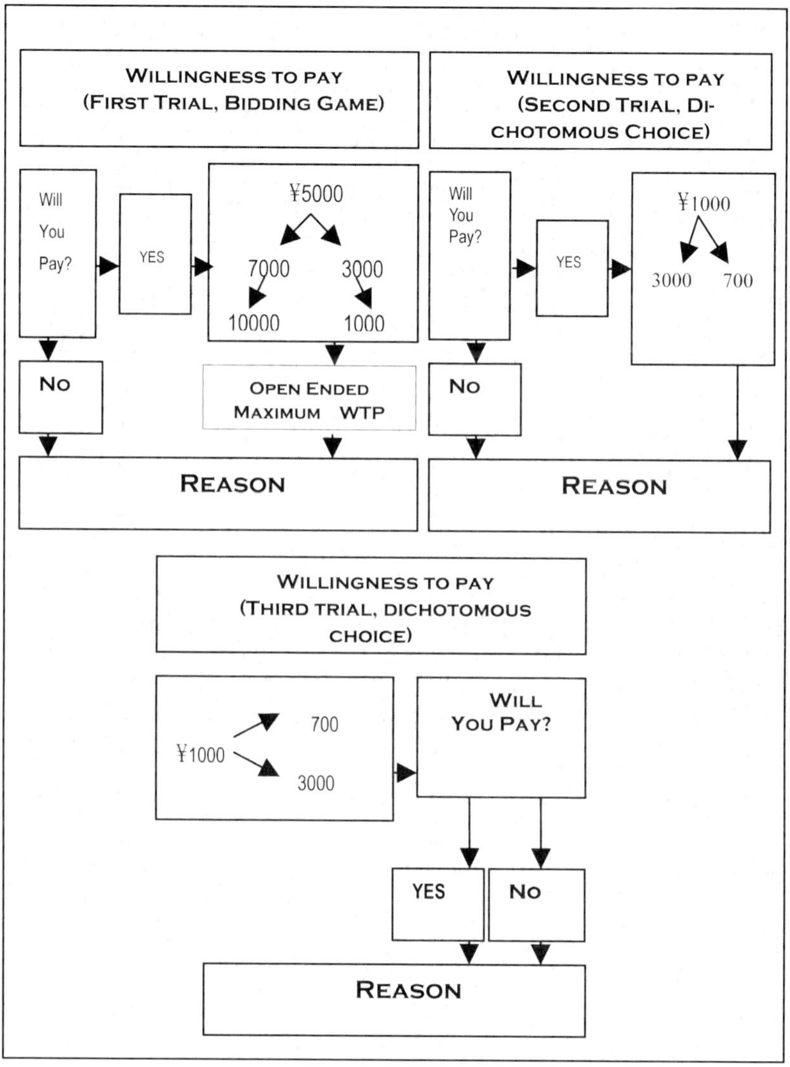

Fig. 4.2. Questionnaire format used in pre-testing study

The age distribution pattern of the respondents showed that the average age is highest in the third trial (approximately 37 years with a standard deviation of 11 years). Thus the third trial had wider range of participants from different age group respondents.

The percentages of the respondents willing to pay were 82 percent in the first trial, 41 percent in the second trial and 53 percent in the third trial (see Table 4.1 and Fig.4.3). The higher percentage of WTP in the first trial may be partially due to the nature of the elicitation method used (bidding game method) and also due to the small sample size. However, the percentages of respondents not willing to pay were highest in the second trial (57 percent).

Table 4.1. Comparison of pre-testing trial results

	First Trial	Second Trial	Third Trial
WTP Elicitation Method used	Bidding Game	Dichotomous choice	Dichotomous choice
Number of Bid Values	One	Three	Three
Number of Respondents	17	42	103
Willing to Pay (YES)	14	17	55
(% YES)	(82)	(41)	(53)
Not Willing to Pay (No)	3	23	37
(% NO)	(17)	(57)	(45)
No Comment	———	1	2
(%)		(2)	(2)
Mean Age	26.6	19.4	37.5
(Standard Deviation)	(10.29)	(0.50)	(11.44)
Mean (WTP)	¥3,559	¥3,470	¥7,561
Median (WTP)	¥1,000	¥1,345	¥3,105

Note: Numerals for some groups are based on less than the full sample because of non-response by some participants.

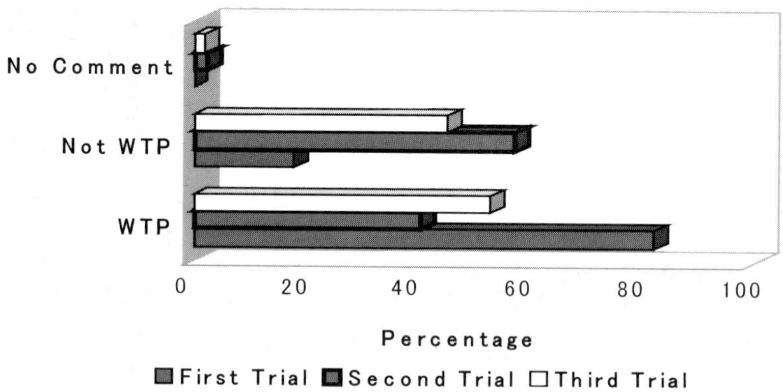

Fig. 4.3. Percentage of willingness to pay

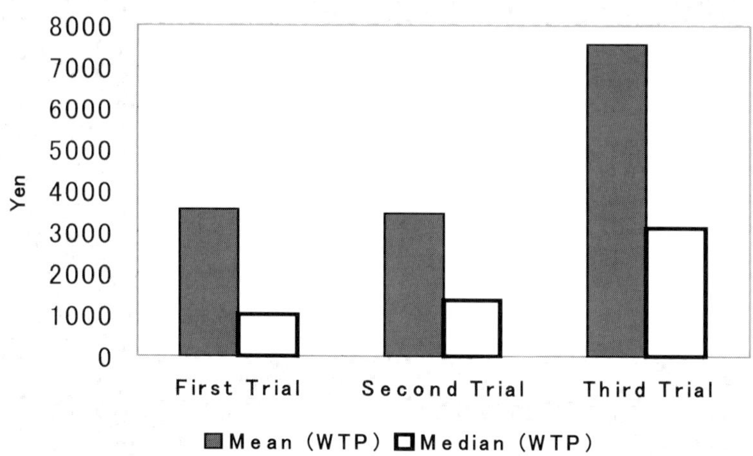

Fig. 4.4. Mean and Median willingness to pay

The reason for this might be attributed to the age concentration of the samples (standard deviation 0.50) and also the placement of the questions regarding the reasons for not willing to pay before the follow-up question. This will be discussed in more detail in the concluding part of the discussion.

Figure 4.4 illustrates the Mean and Median WTP of the three trials. The Mean WTP's are ¥3,559, ¥3,470 and ¥7,561 for the first, second and third trial, respectively. While the Median WTP is very low compared to the mean WTP, both follow the same rising trend. From the findings of the analysis we can conclude that, dichotomous choice (DC) elicitation method provides higher WTP value compared to bidding game (BG) method, even for the same good to be valued. DC method also provides higher WTP compared to open-ended (OE) format. Mitchell and Carson (1993) reported that, the DC/OE willingness to pay ratio generally range between 1.1 and 5 in previous studies. In our study, DC method is selected not for higher WTP but because of its proven efficiency as a balanced elicitation method.

4.4.4 Lessons for Questionnaire Design

1. From the pre-testing of the questionnaire it was found that, while following dichotomous choice technique with follow-up questions, the reasons for not willing to pay should not be asked right after the first bid (see Fig.4.2). Rather the respondent should be given the opportunity to try the next bid, so that he or she can change decision by seeing the next lower bid. Followed by this, the reasons for not willing to pay should be asked. This will help the respondent to think more seriously about the WTP amount in case if the first bid was higher for him/her.
2. While selecting the sample size, a wide range of age group should be covered as far as possible, otherwise concentration of respondents in a particular age group might not give a representative WTP value (see Table 4.1, second trial)
3. The bid levels should be increased and the starting of the bid should be centered on the mean without having significant variance from the mean to reduce error. In second and third trial, we had three levels of bid but finally we changed to six levels of bid value in the original study. Unless good guesses for m (location) and s (scale parameters) are available prior to the beginning of the survey (perhaps from pre-tests), it is advisable to increase the number of different bid values, and choose bid values that cover a relatively broad portion of the range for WTP (FAO 2000).
4. The pre-testing results showed that the socio-demographic questions about the respondent should be placed at the end of the questionnaire. According to Mitchell and Carson (1993), '...questions about the respondent's personal characteristics-the background questions –are best left to the end of the questionnaire, when the respondent is more relaxed about the interview and less likely to take offense at having the interviewer probe into his private life'.

Thus, while designing the questionnaire for CVM study we should be watchful about, 'how to ask the questions, to whom to ask and with what objective'. The choice of the elicitation method and the sequence and contents of the questions are very important, especially for the study of valuing the Isahaya Bay wetland. Because, to some people protection from natural catastrophe, such as flood, is more important than preserving wetland. The subsequent chapters of this study aims to bring these contrasting issues to light.

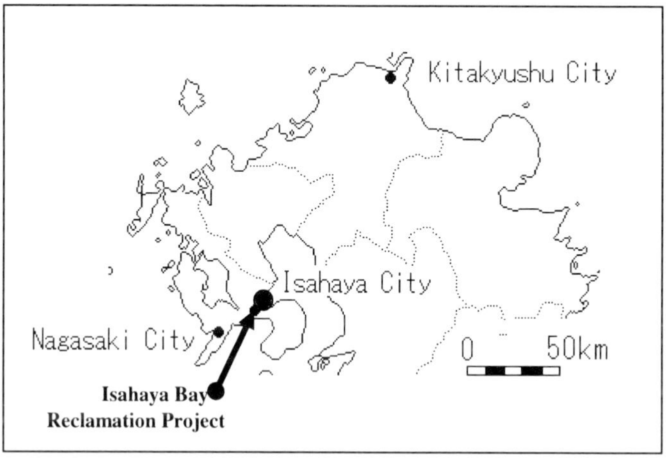

Fig. 4.5. Location of the survey areas

4.5 Survey Description

4.5.1 Study Region and Selection of Respondents

A total of 1,800 questionnaires were distributed in the three cities of Kyushu- Isahaya, Nagasaki and Kitakyushu (600 each), by considering the expenditure and time involved in the survey. The reason for selecting these three cites is to examine the WTP in Isahaya city, where the IBRP is located and also in Nagasaki city, the prefecture that includes the Isahaya city. On the other hand, Kitakyushu, a city located at the topmost part of Kyushu Island, is selected to see the change in WTP due to increase in the distance from the IBRP. The locations of the sample cities are shown in Fig. 4.5. Mail survey technique is used for data collection and households are selected randomly from the registered telephone directory of each targeted city.

4.5.2 Contents of the Questionnaire

The questions contained in the questionnaire of this study can be categorized into three broad headings (see Fig. 4.6). Before getting into the first part, a description is given regarding the problem at hand: the Isahaya Bay Wetland, the Isahaya Bay Reclamation Project, and its effect on the surrounding eco-system, the extent of reclamation and protection; with satellite images showing the changes in the eco-

A. Warm-up questions related to the Isahaya Bay wetland

1. Did you go to visit the Isahaya Bay wetland?
2. Do you know the terms- the Isahaya Bay, Bluespotted Mud-hopper, reclamation and eco-system?
3. Image of the Isahaya Bay reclamation project (IBRP).
4. How IBRP is influencing your life style?
5. How IBRP is affecting you?

B. WTP elicitation questions

6. Would you be willing pay 1,000 yen for protecting the Isahaya Bay wetland?
   (YES→¥3,000, NO→¥700)
   [The starting bid value is changed from ¥1,000 yen to ¥15,000 yen randomly]
   6-1. Reason for willing to pay
   6-2. Reason for not willing to pay
   7-1. Do you know about the Disaster Insurance System (DIS)?
   7-2. Would you enter into the DIS if available?
   7-3. How much you are willing to pay annually, for entering into DIS?

C. Socio-demographic questions

8. Gender
9. Age
10. Profession
11. Annual Income
12. Name of the town and period of stay
13. Participation in volunteer activities
14. Free comments

Fig.4.6. Contents of the questionnaire used in the survey

system due to the initiation of the IBRP. Regarding the picture behind the payment, we have stated that the entire Isahaya Bay would be protected by the proposed fund, and that the Bay would be restored back to the place as it was before the initiation of the reclamation project. Accordingly, the question asked to elicit the respondent's willingness to pay (WTP) is: *Suppose we are going to create a fund named 'Isahaya Bay Wetland Preservation Fund' to restore the eco-system of the IBW. Would you be willing to contribute_______yen (assigned randomly ranging from ¥1,000 to ¥15,000) to this fund only once, to protect the IBW?* This was followed by a follow-up question where the amount is increased or decreased, depending on whether the respondent's initial answer was positive or negative, respectively (see Appendix A for the full questionnaire).

4.5.3 Versions of the Questionnaire

As shown in Table 4.2, the questionnaires were divided into six versions or groups with different amounts of starting bid value under double-bounded dichotomous-choice elicitation method and are distributed on 13 September 2001. The number of versions of the questionnaire were increased to six versions as our pre-testing results revealed that relatively higher number of different versions of questionnaire with different bid values reduce error of the estimation (Gotoh et al. 2002).

4.5.4 Response Rate

The response rates are 21 percent in Isahaya, 20 percent in Nagasaki and 27 percent in Kitakyushu (see Tables 4.3 and 4.4 for details).

4.5.5 Demographic Characteristics

Table 4.3 illustrates the demographic characteristics of the respondents of the study. The respondents of the three cities are relatively closer in terms of average age and income. Comparing the sample demographics with actual one, it was revealed that the sample respondents are slightly older in terms of age and percent of male is relatively higher in some of the cities. But these sampling biases are within statistically tolerable range and are not going to impinge on the results of the study.

Table 4.2. Bid values used under double-bounded dichotomous-choice elicitation method

Versions	Starting Bid Value (Yen)	Second Bid Value	
		YES (Yen)	NO (Yen)
First	1,000	3,000	700
Second	3,000	5,000	1,000
Third	5,000	7,000	3,000
Fourth	7,000	10,000	5,000
Fifth	10,000	15,000	7,000
Sixth	15,000	20,000	10,000

Table 4.3. Questionnaires distributed and response rate

	Isahaya City	Nagasaki City	Kitakyushu City
Questionnaires Distributed	600	600	600
Questionnaires Returned	124	117	160
Response Rate (%)	21	20	27

Table 4.4. Demographic characteristics

	Isahaya City	Nagasaki City	Kitakyushu City
Average Age (Year)	52	51	48
Average Percent Male	56	58	43
Average Annual Income (in ten thousand yen)	400-499	400-499	500-599

4.6 Findings of the study

4.6.1 Image of the Isahaya Bay Reclamation Project (IBRP)

According to the respondents of Nagasaki and Kitakyushu, the strongest image of the IBRP is as a symbol of eco-system (Nagasaki 72 percent and Kitakyushu 77 percent), while to the respondents of Isahaya the strongest image is as a symbol of flood protection (50 percent) (See Fig. 4.7). Thus, the residents of Isahaya regarded IBRP as a protection from flood rather than destruction of natural environment.

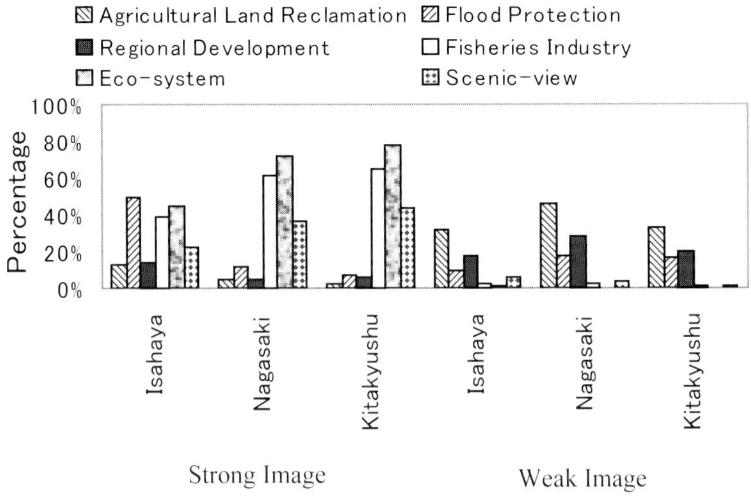

Fig. 4.7. Image of IBRP

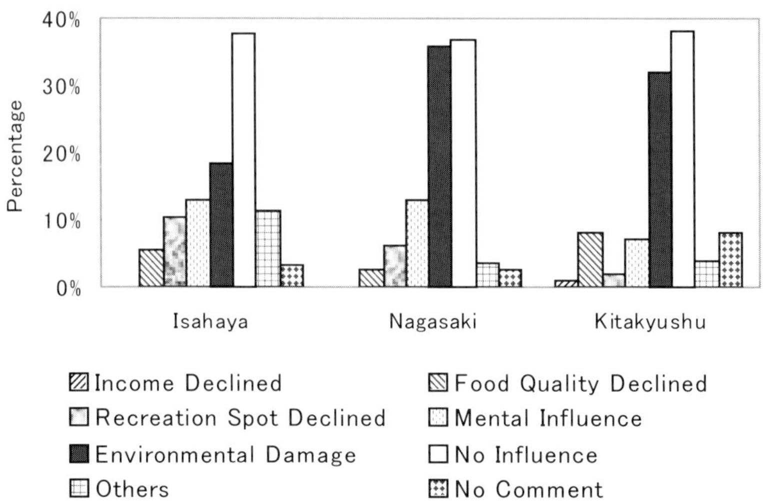

Fig. 4.8. Influence caused by IBRP

4.6.2 Influence of the Project

Regarding the influence of IBRP, 38 percent, 37 percent and 38 percent of the respondents in Isahaya, Nagasaki and Kitakyushu respectively, reported that they had no influence from the project (see Fig. 4.8). Whereas, the second highest influence was reported to be environmental damage. Thus, respondents of all of these three cities, including Isahaya, have recognized IBRP as detrimental to natural environment.

4.6.3 Willingness to Pay (WTP)

Data were analyzed using both Turnbull and Weibull method. The respondents who had resisted to pay in all of the two bids are considered to have zero WTP. However, their opinion regarding refusal to pay were summarized in order to know the reasons for not willing to pay. Based on Turnbull method, the mean WTP's are ¥6,440, ¥6,560 and ¥6,567 for Isahaya, Nagasaki and Kitakyushu, respectively (see Table 4.5). If we multiply these figures by the number of households in each respective city, then we can find that, Isahaya Bay Wetland is worth approximately ¥218 million, ¥1.1 billion and ¥2.7 billion to the residents of Isahaya, Nagasaki and Kitakyushu, respectively (see Table 4.6). This implies the extent of importance of the Isahaya Bay Wetland to the residents of particular city. From the economic point of view, it can be concluded that the utility function of the residents of the each city would increase by this amount of total WTP, if the IBW were restored back to the position as it was before the initiation of the project. The WTP figures under Weibull method are plotted in Fig. 4.9, which shows that the median WTP's of Isahaya, Nagasaki and Kitakyushu as ¥1,282, ¥2,517 and ¥4,108, respectively (see Table 4.7).

Table 4.5. WTP according to Turnbull Method

	Isahaya City	Nagasaki City	Kitakyushu City
Sample Size (complete)	121	115	160
Mean WTP (￥/household)	6,440	6,560	6,567
Median WTP (￥/household)	850	4,000	4,000
SD of the Mean	782.81	0	584.46
Range of 94% confidence interval (￥)	± 1,534.28	-	± 1,145.51

Table 4.6. Total estimated WTP

	Isahaya City	Nagasaki City	Kitakyushu City
Mean WTP (¥/household)	6,440	6,560	6,567
Total Number of Households (in thousand)	34	168	412
Estimated Total WTP (in million yen)[1]	218	1,100	2,706

[1]Estimated Total WTP is calculated by multiplying Mean (WTP) per household by number of household in the respective cities.

Table 4.7. WTP according to Weibull Method

	Isahaya City	Nagasaki City	Kitakyushu City
Mean WTP (¥/household)	6,388	6,524	6,172
Median WTP (¥/household)	1,282	2,517	4,108

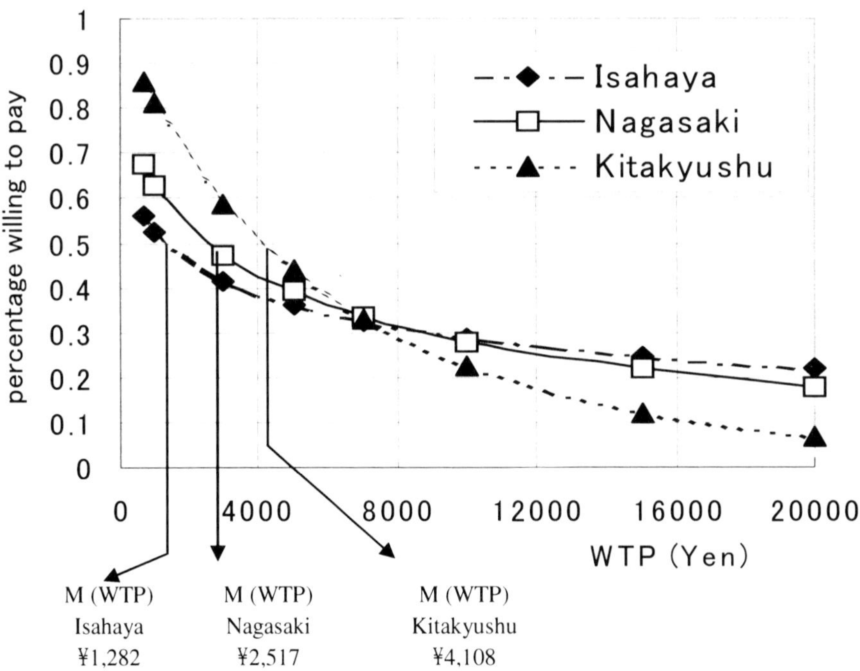

Fig.4.9. Percentage of respondents WTP

4.6.4 Area- wise Willingness to Pay

More specific willingness to pay analysis was also conducted in each city. This would lead to some interesting results. For this purpose, each city is divided into three parts. The particular dichotomous choice (DC) elicitation method used in the study, for each bid value used in the analysis a sample of n=30 or so are needed for statistical significance (Whitehead 2000). Adequate samples were obtained to make calculation as shown in Figs. 4.10 through 4.12. Accordingly, it was found that, in Isahaya the lowest mean WTP is reported from the respondents living near the reclamation project sea dikes, which is ¥3,419 (see Fig. 4.10). Whereas, in central and northern part of the city, the mean WTP is almost doubled to ¥7,124 and¥ 6,377, respectively.

On the other hand, the mean WTP is ¥7,501, ¥7,713 and ¥5,702 in southern, central and northern part of the Nagasaki City respectively, which shows an almost uniform trend (see Fig. 4.11). Finally, in Kitakyushu the same mean WTP is calculated as ¥6,467 at the Yahatanishi ward, ¥7,280 at the Yahatahigashi ward and ¥5,946 at the Kokurakita ward (see Fig. 4.12).

4.6.5 WTP Based on Age

The mean WTP is calculated based on age group of the respondents. For this purpose the whole sample of respondents is divided in three classes (see Fig. 4.13). The results showed that, the middle aged group (i.e., from 40 to 60 years), are willing to pay more in Isahaya and Nagasaki city (¥7,540 and ¥8,474 in Isahaya and Nagasaki, respectively). Whereas, in Kitakyushu City, the willingness to pay amount increases with age of the respondents (mean WTP ¥7,994 for the respondents above 60 years).

4.6.6 WTP Based on Income

On the other hand, mean WTP calculated based on the yearly income of the respondents showed a relatively apparent trend in Isahaya and Nagasaki (see Fig. 4.14). But in Kitakyushu City, although the mean WTP falls from ¥7,069 to ¥6,260 in income group of ¥3 million to¥6 million, it again rises to ¥6,772 for income group of respondents above ¥6 million.

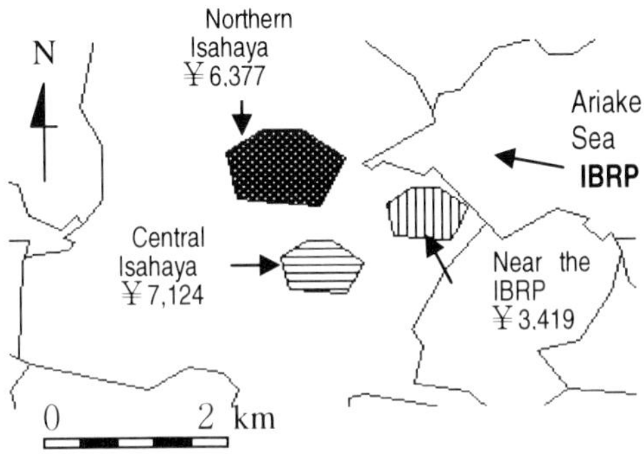

Fig.4.10 Area wise WTP variation in Isahaya City

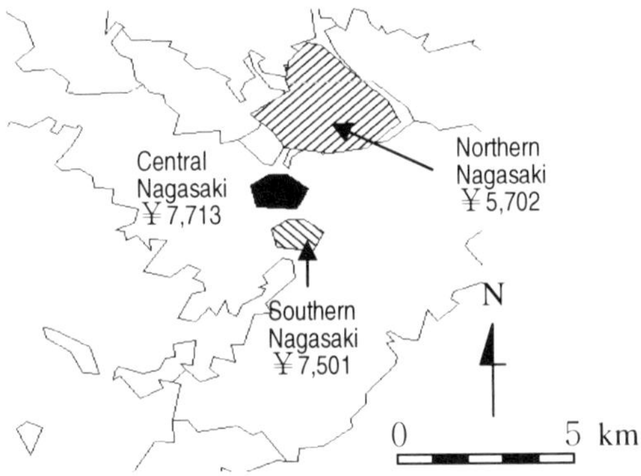

Fig.4.11. Area wise WTP variation in Nagasaki City

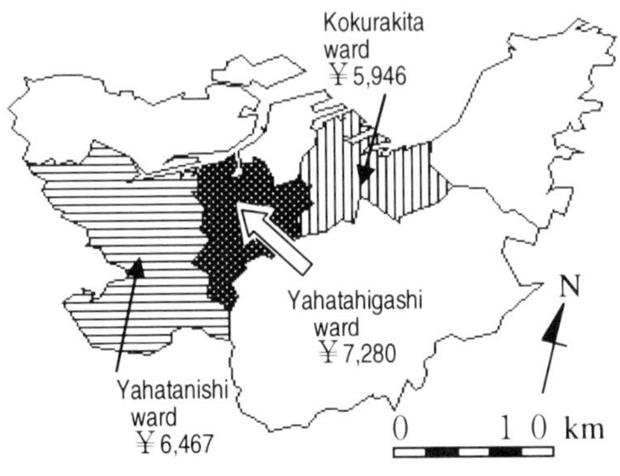

Fig.4.12. Area wise WTP variation in Kitakyushu City

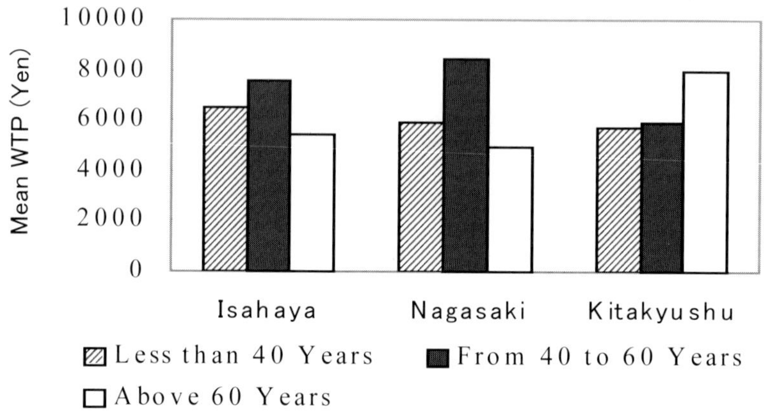

Fig. 4.13. WTP based on age of the respondents

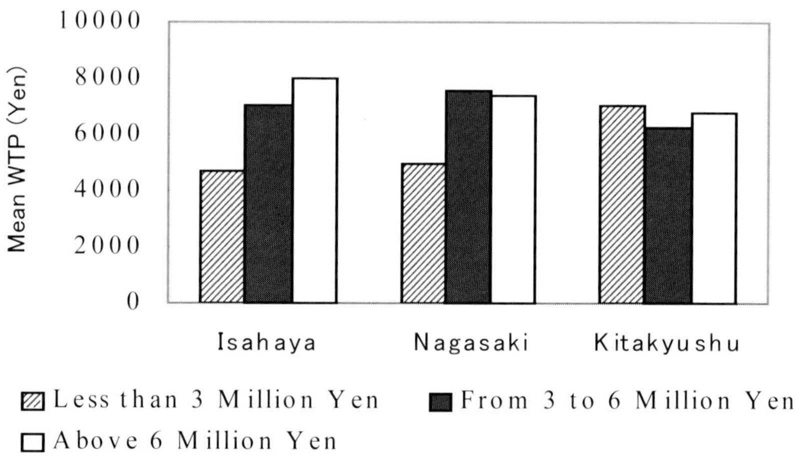

Fig. 4.14. WTP based on income of the respondents

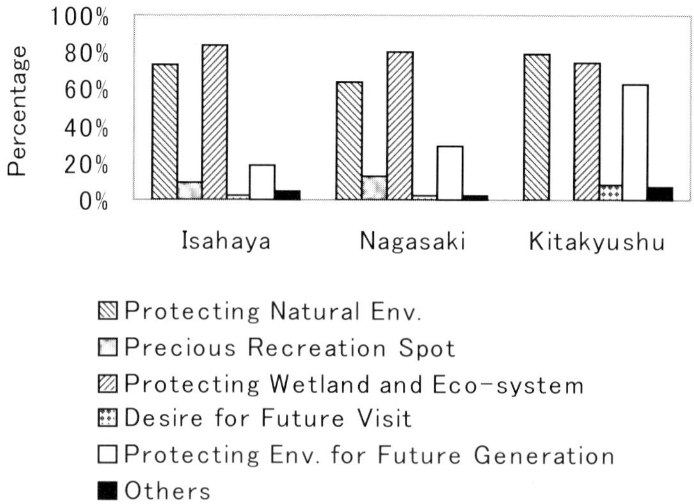

Fig. 4.15. Reasons for willing to pay

4.6.7 Reasons for Willing to Pay

Figure 4.15 presents the reasons why the respondents are willing to pay to restore the Isahaya Bay Wetland. From the figure we can see that, protecting wetland and the eco-system is the most common reason cited by the respondents, as the reason for payment (83 percent, 80 percent and 74 percent in Isahaya, Nagasaki and Kitakyushu respectively). On the other hand, only 10 percent, 13 percent and none in Isahaya, Nagasaki and Kitakyushu, respectively mentioned the reason to pay, by considering the Isahaya Bay Wetland as a recreation spot.

4.6.8 Reasons for Not Willing to Pay

Among the reasons for not willing to pay, 58 percent of the respondents of Isahaya City cited that, agricultural land reclamation and disaster prevention is more important than protecting the environment (see Fig. 4.16). On the other hand, 31 percent, and 51 percent of the respondents in Nagasaki and Kitakyushu respectively, quoted unwillingness to pay as a fund. In Kitakyushu, 37 percent of the respondents reported that the proposed amount is high. This means that, although they cannot afford to pay, but they took the survey seriously.

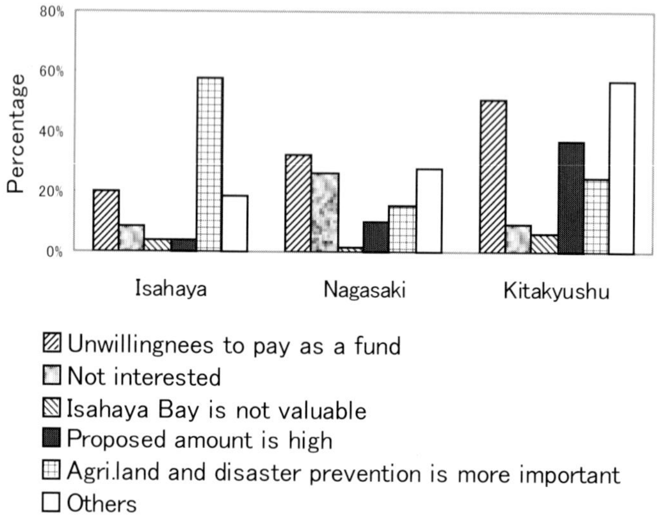

Fig. 4.16. Reasons for not willing to pay

Table 4.8. Actual versus sample demographic characteristics

	Isahaya City		Nagasaki City		Kitakyushu City	
	Actual	Sample	Actual	Sample	Actual	Sample
Age (Year)	40	52	41	51	43	48
Percent Male	47	56	47	58	48	43
Annual Income (in ten thousand yen)		400-499	403	400-499		500-599

Finally, while considering the results of the study we should bear in mind that, respondents showed unwillingness to pay for reasons like: *unwillingness to pay as a fund, agriculture and disaster prevention is more important* etc., also had some willingness to pay if for example *the payment method is changed* or *the advantage of the disaster prevention* was not associated with the IBRP and so on. Thus, by considering their willingness to pay as zero, according to the typical methodology of DC elicitation method, there remains the possibility of underestimating WTP.

Hence, the results reported in this study should be regarded as conservative estimate.

4.7 Biases and Robustness of the Results

The basic assumption is that, persons having interest on public parks have returned the questionnaire. However, we have attempted to determine non-response bias usually arises in CVM studies due to lower response rate. Table 4.8 shows the comparison of the demographic characteristics of the respondents of the study with that of the population characteristics. Comparing the sample demographics with actual one, we have found that the sample respondents are slightly older in terms of age and percent male is relatively higher in some of the cities. But these sampling biases are within statistically tolerable range and are not going to impinge on the results of the study.

Diagnostic tests were run on the analysis to rule out extreme values and ensure the reliability of the results. However, as CVM studies are based on survey data, this study also contains the limitations that survey based studies usually have.

4.8 Summary and Conclusion

A comprehensive study is conducted for determining the value of the Isahaya Bay Wetland in three cities of Kyushu Island, Japan (Isahaya, Nagasaki and Kitakyushu), by applying contingent valuation method. The results lead to the following observations:

1. From the findings it might appear that the eco-system of the Isahaya Bay Wetland is more valuable to the residents of Nagasaki and Kitakyushu than Isahaya (see Table 4.5). But we should take care in interpreting this result and take the following points into account:
    a) The number of households is smaller in Isahaya than in Nagasaki and Kitakyushu, which gives a lower total WTP in spite of approximately the same average of WTP as other cities (see Table 4.6).
    b) To the residents of Isahaya, protection from flood disaster is more important than the value of wetland, which was also reflected into the image survey question where 50 percent of the respondents of Isahaya strongly viewed the IBRP as a symbol of flood protection (see Fig. 4.7).
2. The gradual rising trend in median WTP as shown in Fig. 4.9 indicates that, as we go more further from the IBRP, more the median WTP increases. Thus, it can be concluded that, in case of the particular commodity like the Isahaya Bay Wetland, where sea dikes are constructed having both positive (flood control)

and negative (environmental damage) impact, the further we go from the location of the project, more the median WTP increases.

3. From Figs. 4.10, 4.11 and 4.12 we noticed the area-wise variation in mean WTP for the three cities. In Isahaya City especially this value is significantly lower as reported by the respondents living near the reclamation project. Interestingly this value increases by two-fold in the areas a little far from the project in the same Isahaya city, such as central or northern Isahaya. This supports the second observation made above and leads us the conclusion that, the mean WTP also increases as we go far from the location of the project and after a reasonable distance the amount become steady gradually.
4. There is a sizeable gap between median WTP and mean WTP, a difference that has large implication on policy makers concerning the appropriateness of continuing the IBRP.
5. It can also be generally concluded from the results of the analysis that, willingness to pay for restoring the eco-system increases with age and income in the sample cities, showing the higher propensity of willingness to pay by the middle aged and income group of respondents.
6. Although the respondents of Isahaya city view IBRP more as protection from flood (50 percent) than damaging eco-system (44 percent), but their mean WTP is very close to those of Nagasaki and Kitakyushu (see Table 4.5). This implies that, they are supporting the project for protecting them from flood disaster; at the same time they also acknowledge the destruction of eco-system caused by the project.
7. The results of the study have several implications for the society and policymakers as follows:
    a) Firstly, the total WTP calculated can be treated as a proxy value of the amount of damages caused to the residents by the IBRP (see Table 4.5 for the Total WTP).
    b) Secondly, these WTP values calculated by using the contingent valuation method can provide a base for paying compensation to the potential affected interested groups by the concerned authorities.

In the next chapter we are going to see how the valuation of the IBW by applying CVM affects the cost-benefit analysis of the IBRP.

CHAPTER 5 Recalculating Cost-Benefit Analysis of the IBRP

5.1 Introduction

In Chapter 4 the value of the Isahaya Bay Wetland (IBW) is estimated by applying contingent valuation method (CVM). Accordingly, the objective of this chapter is to re-calculate the cost-benefit analysis of the Isahaya Bay Reclamation Project (IBRP) and to suggest ways to incorporate socio-environmental considerations in the cost-benefit analysis of a particular project.

5.2 Cost-Benefit Analysis of the IBRP

Table 5.1 shows the economic impact calculation of the Isahaya Bay Land Reclamation project as projected by the Ministry of Agriculture, Forestry and Fisheries (MAFF) of Japan. This table shows the yearly total economic benefits of the project as ¥8,512 million and considered only maintenance cost of ¥145 million while calculating the net benefit of the project. And Table 5.2 illustrates the cost-benefit analysis (CBA) of the project, which calculates the present value of the total benefits of the project as ¥138,500 million and present value of the cost of the project as ¥135,000 million. This makes the project viable, as the benefit-cost (B/C) ratio is greater than one (1.026>1).

Table 5.3 shows the recalculation of the CBA of the same project in the year 1996, as the initially estimated project cost rises due to the increase of raw material cost and change in project plan (Miyairi 1998). This makes the project unviable, since the B/C ratio is less than one (0.584<1). Hence, by reviewing the CBA of the IBRP critically, several drawbacks can be drawn as follows:

1. Only considers maintenance and construction costs as the cost of the project.
2. Did not consider the impact of the project on fishing industry and local industries supported by the fishing resources of the region.
3. Did not consider the economic value of the tidal flat wetland composed of high organic matter that might be lost due to the initiation of the project.
4. Ignored impact of the project on the life style of the local residents.

5. Did not consider the high economic value of the Isahaya Bay Wetland as sightseeing resources.
6. Absence of sensitivity analysis to investigate the robustness of net benefit estimates to different degrees of uncertainty.

Table 5.1. Economic Impact Assessment of the IBRP (In million yen)

Category	Amount/ Percentage	
1. Crop production benefits	2,640	31.0
2. Land use benefits	1,478	17.4
3. Disaster protection benefits	4,040	47.5
3.Contribution to Maintenance cost	-145	-1.7
4. Others	499	5.9
Total	8,512	100.0

Source: National Reclamation Project: The Isahaya Bay Region Total Execution Plan, Kyushu Regional, Agricultural Administration Office, Ministry of Agriculture, Fisheries and Forestry (MAFF), 1987, pp.28-34

Table 5.2. Cost-benefit Analysis of the IBRP in 1986

Category	Amount/ Ratio
1. Net yearly benefits	¥8,512 million
2. Capital recovery factor	0.06
3. Present value of total benefits	¥138,500 million
4. Present value of total project costs	¥135,000 million
5. Benefit-cost ratio (3/4)	1.026

Source: Same as Table 1

Table 5.3. Cost-benefit Analysis of the IBRP in 1996

Category	Amount/ Ratio
1. Net yearly benefits	¥8,512 million
2. Capital recovery factor	0.06
3. Present value of total benefits	¥138,500 million
4. Present value of total project costs	¥237,000 million
5. Benefit-cost ratio (3/4)	0.584

Source: Recalculated from Table 2 (Miyairi, 1998)

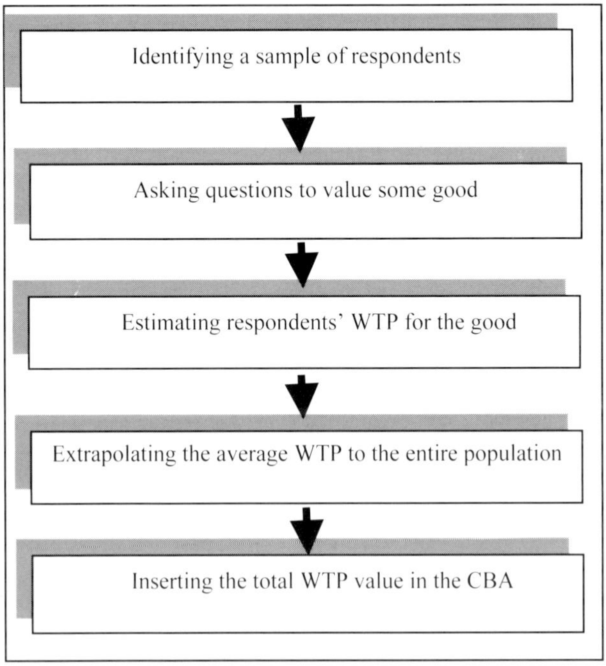

Fig. 5.1. Steps involved in CBA using CV methods

These shortcomings of the cost-benefit analysis of the IBRP are very significant, and hence, strongly criticized by different research studies. Miyairi (1998) observed that, '.....considering only the positive impacts of a project and ignoring the negative impacts like the IBRP might be termed as converting things to one's own advantage'.

However, these negative impacts of public construction projects, which mainly refers to socio- environmental impacts (e.g., impact on fishing industry, wetland, life-style of residents, eco-tourism spot etc.), might be included in the CBA by using *contingent valuation* surveys which can elicit individual's preferences regarding a particular public good through questionnaire survey.

In this part of the study we took a very modest attempt to recalculate the cost-benefit analysis of the IBRP by inserting the opinion of the residents living in and around Isahaya city in monetary figures, regarding the impact of the IBRP on their life style and surrounding environment.

5.3 Methods

In case of public goods like the Isahaya Bay Wetland, where the IBRP is located, there are simply no or very poor market proxies or other means of inferring about the preferences of the residents. This leaves us with no alternative but to ask the

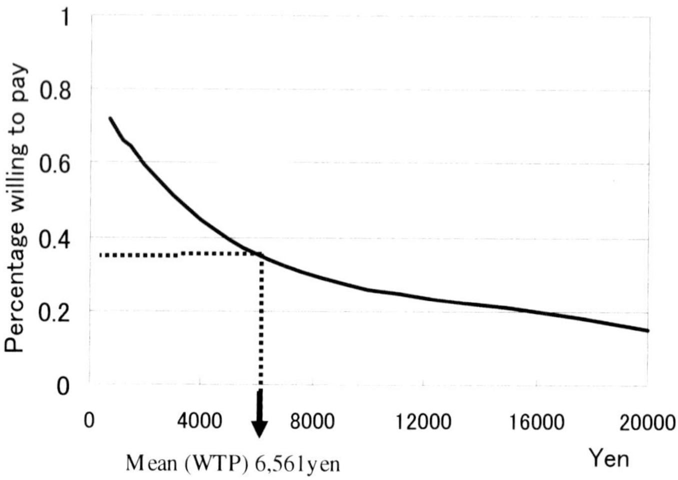

Fig. 5.2. Percentage of respondents willingness to pay

residents about their opinion regarding the impact of the IBRP under Contingent Valuation survey questionnaire. In cost-benefit analysis (CBA), questionnaires designed to elicit preferences are normally referred to as contingent valuation (CV) surveys, or sometimes hypothetical valuation surveys, because respondents are not actually required to pay their valuations of the goods (Boardman et al. 2001).

Extracting the public opinion using CV methods and using them in CBA involves the steps as shown in Fig. 5.1. As CV surveys are expensive to conduct, analysts extrapolate the results of existing surveys to different population for drawing conclusions (Boardman et al. 2001).

In estimating the mean WTP, the Turnbull method of non-parametric estimation is followed as described in the previous chapter (see chapter 4 for detailed methodology). Also the database used in this section is also the same as in the previous chapter, with the only difference that here we have accumulated all the individual city data under one set and estimated the aggregate WTP for all the three cities.

5.4 Deriving Total Environmental Benefits of IBRP

On the basis of the methodology described above, the mean (WTP) is estimated to be approximately ¥6,561 per household for all the respondents of Isahaya, Nagasaki and Kitakyushu. (See Table 5.4). The median (WTP) is ¥4,000 per household. In this study, the mean (WTP) is used instead of median (WTP), for calculating the aggregate WTP. This is because, under Turnbull method, mean (WTP) can be calculated in exact figure, unlike median (WTP) that gives only a range of amount.

Figure 5.2 demonstrates the percentage of respondents' willingness to pay at various amounts for restoring the socio-environmental condition of the Isahaya Bay Wetland. An apparent negative sloped curve was obtained on the percentage of willingness to pay with the increase in amounts to pay.

5.5 Recalculating Benefits and Costs of IBRP

The results of the CV survey study indicated that the residents of the three cities of Kyushu- Isahaya, Nagasaki and Kitakyushu are willing to pay ¥5,561 on an average to restore the socio-environmental state of the Isahaya Bay Wetland as it was before the construction of IBRP. Now in order to calculate the total benefits of restoring the Isahaya Bay Wetland, we need to derive the aggregate value by extrapolating the individual average WTP. For calculating this aggregate value, first, we need to define the relevant population; and for determining the relevant population we need to again define the relationship between beneficiaries of the Isahaya Bay Wetland, and those who pay for it. The logic behind this is explained through Fig. 5.3 and Table 5.5.

The procedure to determine the relevant population are explained in Fig. 5.3. Residents living near the Isahaya Bay Wetland would benefit from its preservation, and fall in category I and II. Whereas, residents living far away from IBW fall in category III and IV. Category I represents the ideal population for a CVM study. However non-users should be included too, as they might also have existence and option values. Accordingly, the relevant population is defined as the entire households of Kyushu. The entire household number of Kyushu was used to calculate the total WTP for two reasons: firstly, the IBRP is located in Kyushu and are reported to have adverse impact on the eco-system of different surrounding

Table 5.4. Total estimated WTP

Description	Number/ Amount
Sample size (complete)	396
Mean WTP (¥/household)	6,561
Median WTP (¥/household)	4,000
SD of the Mean	399.3
Range of 94% confidence interval (¥)	± 1,565.3
Total Number of Households in Kyushu	4,961,000
Estimated Total WTP (Million yen)	32,549

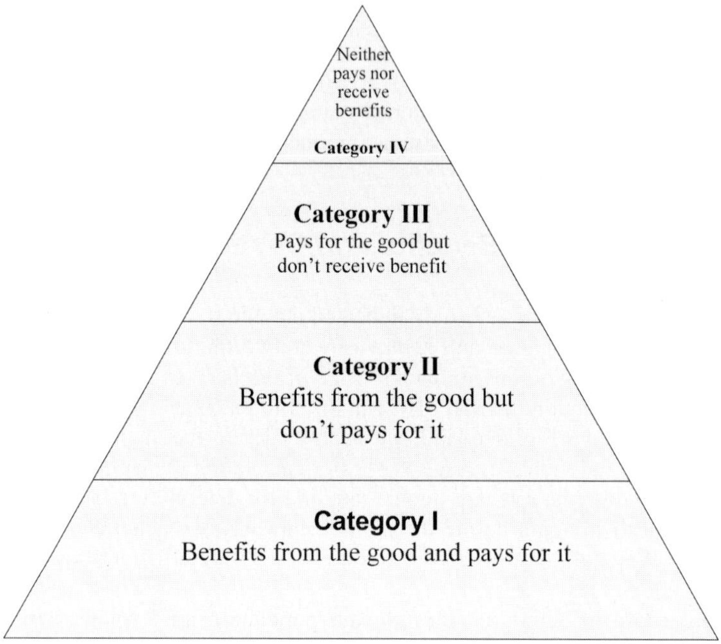

Fig. 5.3. Determining the relevant population

Table 5.5. Relationship between beneficiaries of and payers for an environmental good

		Pays for the Good?	
		Yes	No
Benefits from the Good?	Yes	I	II
	No	III	IV

Source: Mitchell and Carson (1989)

prefectures of Kyushu e.g., Nagasaki, Saga, Kumamoto and Fukuoka, i.e., use value; secondly, the Isahaya Bay Wetland is an important recreational and ecological spot to the entire population of Kyushu as evident from the survey results and, thus, have non-use value to the residents of other areas of Kyushu.

Therefore, after extrapolating the mean WTP of ¥6,561 to the entire 4,961,000 (MPHP 2002) private households of Kyushu it was found that, the residents of Kyushu are willing to pay approximately ¥32.55 billion to restore the Isahaya Bay Wetland (See Table 5.6). This amount might be considered as the cost of IBRP, as the respondent's are willing to pay this amount to restore the ecological condition of the region to the position it was before the initiation of the project.

Table 5.6. Recalculation of cost-benefit analysis of the IBRP

Category	Amount (Million Yen) / Ratio
1. Net yearly benefits	8,512
2. Capital recovery factor	0.06
3. Present value of total benefits	138,500
4. Present value of total project costs	135,000
5. Present value of socio-environmental costs	32,549
6. PV of total project cost (4+5)	167,549

Consequently, the benefit-cost ratio of the IBRP is recalculated in Table 5.6, which is a continuation of Table 5.2, with the same cost and benefit amounts. The only difference between the two is that, here ¥32,549 million (¥32.55 billion) calculated above is inserted as the socio-environmental cost of the project as an additional to the cost of the project. Inclusion of such cost increases the total cost of the project to ¥167,549 million (See Table 5.6). This interestingly revealed that, if environmental costs of the project were considered by such CV survey studies, before initiating the project, then at that time the IBRP was unviable as B/C ratio is less than one (0.827<1). Therefore, if socio-environmental factors were considered, IBRP would have been proved to be unviable even in 1986.

5.6 Summary and Conclusion

In this chapter, we tried to suggest ways to incorporate socio-environmental factors into cost-benefit analysis of the public works project by taking the IBRP as one of the case. The findings can be summarized and concluded through the following points:

1) The IBRP is regarded as a protection from flood to the residents of the Isahaya city, residing near the project spot. However, to the majority of the respondents of Nagasaki and Kitakyushu city and also to a significant portion of the Isahaya city, IBRP is acknowledged as a major ecological spot.
2) Recognition of socio-environmental cost into cost-benefit analysis is very important to let citizens and the concerned public body to decide on the viability of the project.
3) Contingent valuation method (CVM) is recommended as one of the alternatives to incorporate socio-environmental considerations into the cost-benefit analysis.
4) Sensitivity analysis should be included in the cost-benefit evaluation process to judge the validity of the project in different degrees of uncertainty. As it was observed in the case of IBRP, various predictions were made about the future, which later did not remained constant, such as: price of the construction materials, project plan etc.

5) In case of public works project lower B/C ratio does not necessarily imply that the project should be rejected, as some projects urges the need to be implemented in spite of lower B/C ratio. In such a case, socio-environmental costs calculated by using the contingent valuation method can provide a base for paying compensation to the potential affected interested groups.

5.7 IBW Case Study Conclusion

To the best of our knowledge, this study is the first attempt to quantify the environmental value of the eco-system of the Isahaya Bay Wetland (IBW). It has drawn attention to the contrasting views that are often involved in environmental issues, where we need to choose between environmental preservation versus protection from natural calamity. The results of this study would provide the residents and researchers a quantitative estimate to base their discussion about the much-debated IBW.

As suggested in this chapter, recognition of socio-environmental costs of the public works project by using CVM might be considered as one of the alternatives for inserting public opinion in the project evaluation process. This would enable the citizens and the concerned public bodies to decide on the viability of the project before initiating it and reduce unwanted debates.

Regarding the IBRP it can be concluded that, undertaking construction works to protect from natural disaster is definitely important, but careful long-term viability study should must be conducted to make sure that it does not eventually destroy the invaluable and non-restorable eco-system. This area leaves the scope for further studies.

PART III
CASE STUDY 2: PUBLIC PARKS

CHAPTER 6 Estimation of the WTP for Preserving Public Parks

6.1 Introduction

Public parks provide different benefits to the community such as: recreational benefits, economic benefits, environmental benefits, safety benefits and so on. Among them environmental benefits provided by public parks are stressed strongly, as parks significantly improve surrounding environment by contributing to the reduction of heat island phenomenon by increasing greenery, reducing air, water, and noise pollution, and helping in wildlife preservation. Also, recently the Japanese Ministry of Land, Infrastructure and Transport underscored the need for public park maintenance and development, by including it as one of the key agenda amongst thirteen targets set for future public works projects in Japan (NNS 2003). In this context, it is interesting to investigate the state and condition of public parks in Japan. In doing so, Nagasaki City is considered as a case study.

Accordingly, the objective of this chapter is to value the public parks in Nagasaki City by estimating the willingness to pay (WTP) for preserving it, which will provide us a monetary estimate about its importance. And the more specific objectives of the study are as follows:

1. To identify the explanatory factors influencing the willingness to pay.
2. To trace out the main problems prevalent in public parks.
3. To suggest ways to concerned bodies for improvement and development of public parks.

6.2 Public Parks in Nagasaki City

In the ranking list of easily livable urban cities, Nagasaki is ranked as 497th C grade city among 678 urban cities of Japan, where the rank C denotes the fifth rank among seven levels of ranking (Toyo Keizai 2003). For an exotic city like Nagasaki, having long traditions of international linkage, this ranking is very unfortunate. Among the ranking criteria, one of the important factors is the ratio of per person square meter (PPSM) of public parks. This ratio of PPSM is

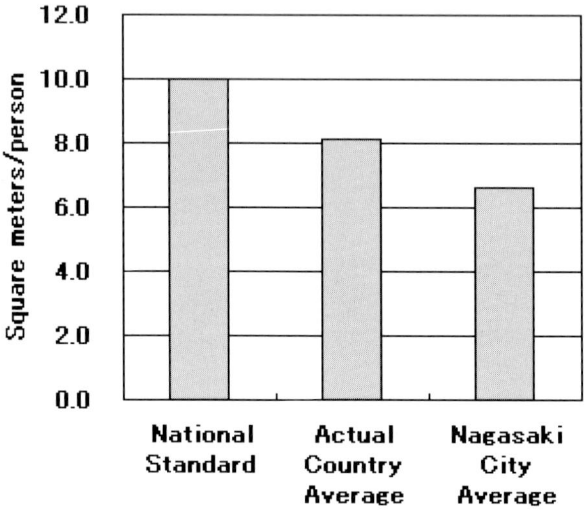

Fig 6.1. A comparison of square meters of public parks per person

alarmingly low in Nagasaki City as compared to the national standard and actual national average of Japan. As we can see in Fig. 6.1, national standard of public park is 9.5 PPSM and the actual national average is 8.1 PPSM. Whereas, in Nagasaki city it is 6.6 PPSM. Thus, although being a hillside city where public parks have more wide uses, adequate space or number of public parks are lacking in Nagasaki City, which can be considered as one of the important reasons for making the Nagasaki city poorly ranked urban city.

Now, on the basis of the background portrayed above, it is interesting to investigate how the residents of Nagasaki City are viewing the need for public parks, in order to plan for the increase and improvement of it. And one common and more easily understood method to articulate the need of some public good, is to express it in monetary terms by valuing it. Consequently, we are going to estimate the willingness to pay for preserving the nearest public parks as a means to express the urge for the maintenance and development of public parks in Nagasaki City. As a method for estimating the WTP, we are going to use contingent valuation method (CVM). The results of the study are expected to provide some concrete guidelines to the concerned interest groups in deciding about the development and/or increase of public parks in Nagasaki City based on empirical study results.

6.3 Justification for Using CVM

In valuing environmental goods like public parks, the value includes both use (value derived from actual use of a good or service)and non-use (also referred to as passive use values, are those not associated with actual use, or even the option to use a good or service) values, because there are people who regularly go to parks and also there are a large group of people who do not go to parks, but feels the need of it and want to pay for it. Accordingly, contingent valuation method (CVM) would be relatively more appropriate for our study, which is used to estimate both use and non-use values and can be applied to find economic values for almost all kinds of environmental amenities. Since non-use values are also significant, other methods, such as the travel cost method, would underestimate the benefits of preserving the public parks.

6.4 Methods

6.4.1 Overview

Double bounded dichotomous-choice (DC) elicitation method has been used for deriving the WTP figures. For calculating the mean WTP, non-parametric estimation Turnbull method is followed, which uses the same Equation 4.1 to calculate the WTP (Turnbull 1976). The estimated mean WTP is extrapolated to the total number of households in Nagasaki City to calculate the aggregate willingness to pay of the residents of Nagasaki City for conserving public parks. Then, a multivariate analysis is conducted by forming an econometric model, to check the variables having influence on WTP. Finally, the results of the multivariate analysis is verified by running diagnostic tests for identifying the existence of outliers, if any and ensuring the robustness of the full model.

6.4.2 Survey Description

a) Study region and selection of respondents

The questionnaire was sent to a random sample of 1,000 households of Nagasaki City. The distribution of the sample respondents is shown in Fig. 6.2. Mail survey technique was used for data collection and households are selected randomly from the registered telephone directory.

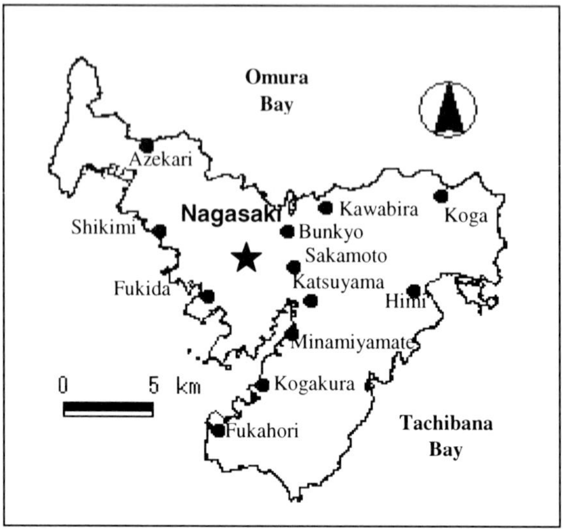

Fig.6.2. Survey area in Nagasaki City

b) Pre-testing study

A Pre-testing study was conducted before entering into the final survey. A total of 23 respondents of different age and profession were questioned to test the survey instrument. Bid values were varied between ¥1000 to ¥12,000 and dichotomous-choice elicitation method was used. In the pre-testing study, the mean WTP were estimated to be approximately ¥4,077. Based on the experience of the pre-testing study, the final survey questionnaire was modified to some extent and the numbers of bid values were finally fixed to five.

c) Contents of the questionnaire

The questions contained in the questionnaire of the study can be categorized into four broad headings. The starting bid value is changed from ¥2,000 to ¥10,000 divided into five different bid values, which were randomly assigned to survey respondents. The contents of the questionnaire and the bid values are summarized in Tables 6.1 and 6.2.

Table 6.1. Contents of the questionnaire used in the survey

Category of questions	Content of the questions
A Current status of public parks	A detailed description regarding the possible and potential use of public parks, their current status in Nagasaki in comparison with the national standard of Japan were elaborated, with the aid of pictures, graphs and supporting data.
B Warm-up questions regarding the image of public parks	1. How far do you know about the possible use of public parks? 2. What kind of place is a public park to you? 3. How many times in a month do you go to public parks and the reasons for going (if you go) and the reasons for not going (if you don't go)?
C WTP elicitation questions	4. Suppose the closest public park to your residence is going to be disappeared for some reason. In order to protect and preserve the parks, we are going to establish a 'Public Parks Protection Fund'. Would you be willing pay --------- yen in this fund, for preserving the public parks in Nagasaki city? (This was followed by a follow-up question where the amount is increased or decreased, depending on whether the respondent's initial answer was positive or negative, respectively). 4-1. Reason for willing to pay 4-2. Reason for not willing to pay
D Socio-demographic questions	5. Gender and family member 6. Age 7. Profession 8. Name of the town, period of stay, and name of the closest public park 9. Annual Income 10. Free comments

Table 6.2. Bid values used

Versions	Starting bid value (Yen)	Second bid value	
		YES (Yen)	NO (Yen)
First	2,000	4,000	1,000
Second	4,000	6,000	2,000
Third	6,000	8,000	4,000
Fourth	8,000	10,000	6,000
Fifth	10,000	12,000	8,000

Table 6.3. Actual versus study demographic characteristics

	Nagasaki City	This Study
Average Age (Year)	42.4	58.9
Average Percent Male	47	49
Average Annual Income (in ten thousand yen)	403	488

d) Response rate and non-response bias

The response rate was approximately 20 percent. Based on the experience of the mail survey response rate of studies conducted in Nagasaki, Japan, the response rate in our study can be considered to be satisfactory for conducting the CVM study (Gotoh et al. 2002; Ahmed, 2002a). Because, except for direct interview, the response rate in mail survey is usually very low. The basic assumption is that, persons having interest on public parks have returned the questionnaire. However, we have attempted to determine non-response bias usually arises from lower response rate and we have found no such possibility by comparing the means of the survey demographic variables with that of the actual data of the Nagasaki City (see Table 6.3). As we can see from the table, that the age, percent male and annual income of the respondents of the study are a little bit higher from that of the actual data of the Nagasaki City, but this difference is not so significant. From this we can conclude that, the sample of 194 is representative of the population we sampled from.

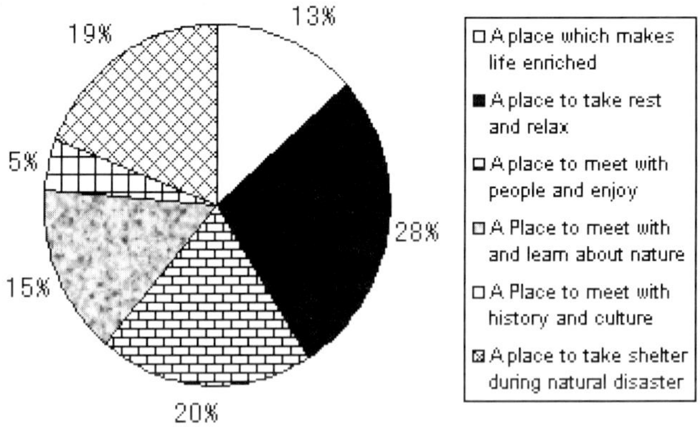

Fig.6.3. Image of public parks in Nagasaki City

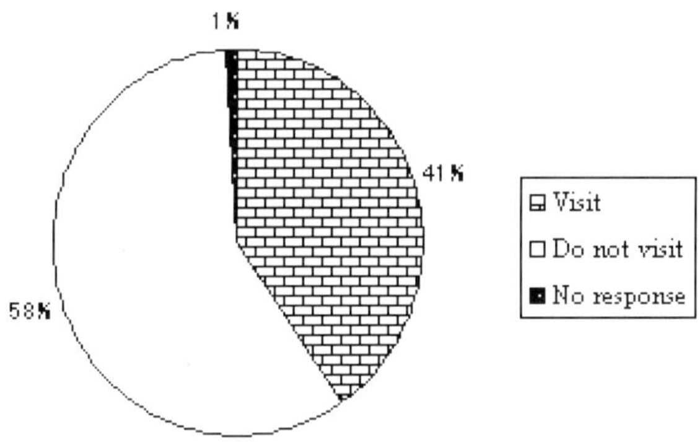

Fig.6.4. Visit rate of respondents to the nearest public parks

6.5 Results

6.5.1 Image of Public Parks

Regarding the image of public parks to the respondents of Nagasaki City, 28 percent and 20 percent of the respondents reported to be a place to rest and relax and also a place to meet with people respectively (see Fig. 6.3). But image of parks as a place to meet and learn about nature were comparatively lower (15 percent).

Thus there are possibilities of lack of presence of nature in public parks of Nagasaki.

6.5.2 Visit in public parks

In response to the question '*whether the respondent visits parks at least once in a month*', 41 percent of the respondents reported that they visit, 58 percent responded that they do not and 1 percent refrained from answering (see Fig. 6.4). This indicates that a large portion of the citizens of Nagasaki City is not visiting parks regularly.

6.5.3 Average Willingness to Pay

As shown in Table 6.4, by applying the non-parametric estimation under Turnbull method cited above, the mean willingness to pay (WTP) for preserving public parks in Nagasaki City is estimated to be ¥5,225 per household. This indicates that on an average each sampled household in Nagasaki is willing to pay ¥5,225 for preserving the nearest public park. If we plot the mean WTP in percentage form against different amount of WTP, then the WTP curve takes the decreasing slope form as shown in Fig. 6.5. From this graph, the gradual decrease in the percentage WTP with corresponding increase in the amount can be observed.

6.5.4Benefit-cost Analysis

In order to conduct CBA we extrapolated the mean WTP to the entire household of Nagasaki City, then the economic value for preserving public parks in Nagasaki City would become ￥920 million (see Table 6.4).

Table 6.4. Willingness to pay

1. Sample Size (complete)	194
2. Mean WTP (￥/household)	5,225
3. Standard Deviation of the Mean	320
4. Range of 94% confidence interval (￥)	±628
5. Total Number of Households*	176,110
6. Estimated Total WTP (in million yen) (Row 2 multiplied by row 5)	920

* Statistical Yearbook of Japan, MPHP 2002

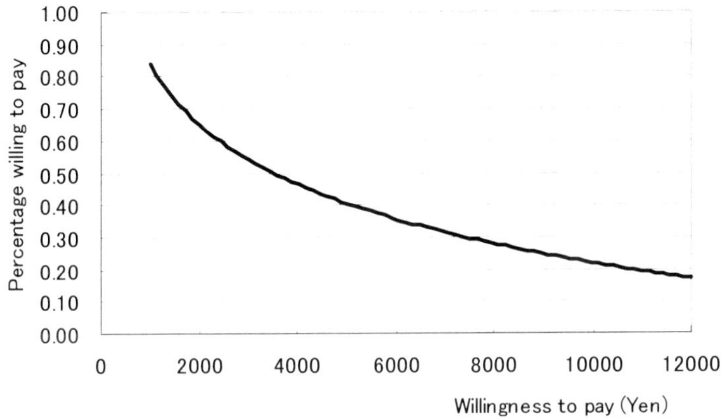

Fig.6.5. Percentage of respondents WTP

6.5.5 Reasons for Willing to Pay

Figure 6.6 presents the reasons why the respondents are willing to pay to protect the public parks in Nagasaki City. From the figure we can see that, the majority of the respondents want to pay for preserving parks for the family, society and the nature (23 percent, 20 percent and 17 percent, respectively). On the other hand, only 4 percent mentioned that the proposed bid value was low as a reason for willing to pay. Thus the economic value of preserving public parks calculated in this study contains a big amount from the feeling that the parks are also needed for the future generation. And the influence of the proposed amount on the respondents who agreed to pay was insignificant.

6.5.6 Reasons for Not Willing to Pay

Among the reasons for not willing to pay, 31 percent of the respondents quoted that the *proposed amount is high* (see Fig. 6.7). The next major reasons cited are *people using public parks are few* and *lack of nature in public parks* (22 percent and 17 percent, respectively). Thus, absence of nature and non-visit to public parks seemed to be related, as was also mentioned as one of the main reasons for not going to parks (see Fig. 6.3).

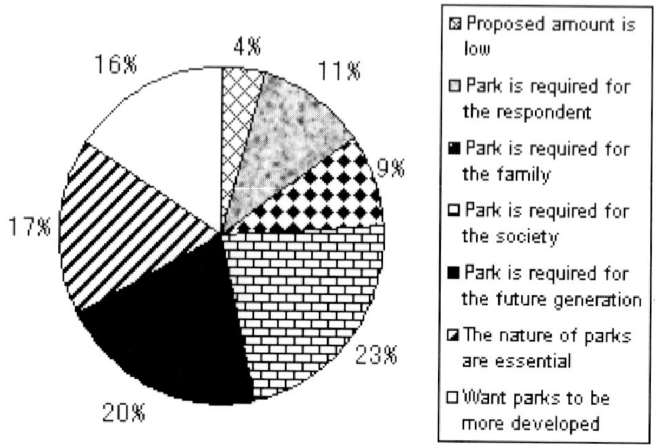

Fig.6.6. Reasons for willing to pay

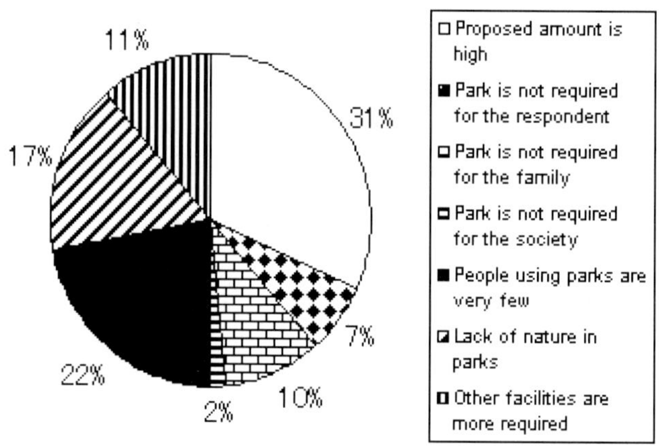

Fig.6.7. Reasons for not willing to pay

6.6 Multivariate Analysis

A multivariate model is constructed to determine the factors, which influence willingness to pay for preserving public parks in the Nagasaki City. It comprises the variables that are assumed to have an influence on willingness to pay, such as: bid value, visit, income and demographics. Accordingly, bid value and income are included to test the *rational choice theory*. Visit is included to test the impact of *non-use value*. And finally, demographic variables such as age and sex are included to check whether *social forces* plays a role in the willingness to pay. These factors, which are hypothesized to have influence on WTP based on economic theory, can be summarized by the following conceptual model:

$$\text{WTP} = f(\text{bid, visit, income, demographics}) \quad (6.1)$$

In case of CV studies with dichotomous structure of the dependent variable, the model can be estimated through a non-linear probability distribution and the most common one is the logit model (Pate and Loomis 1997). Because of the discrete nature of the dependent variable, the ordinary least square regression can be used to fit a linear probability model, but its heteroskedastic nature of the error terms and the possibility to predict values beyond the range of 0 and 1, led us to use the logistic regression model. Accordingly, the logistic regression model, which can be developed to analyze hypothesized model in Eq. (6.1), takes the following shape:

$$\ln[\text{prob(yes)}/\{1-\text{prob(yes)}\}] = b_o - b_1\,(\text{lbid}) + b_2\,(\text{visit}) + b_3\,(\text{income}) + b_4(\text{age}) + b_5(\text{sex}) \quad (6.2)$$

The explanation of the variables and the descriptive statistics are presented in Table 6.5. Thus, the willingness to pay is expected to be inversely related with the bid value, positively with the number of visit and the ability to pay (income).

6.6.1 Results of the Multivariate Analysis

Logistic regression results are presented in Table 6.6 and the analysis is conducted by SPSS. The results of the model indicate that, the coefficient of lbid,

Table 6.5. Variables and Descriptive Statistics (n =194)

Variable Name	Variable Description	Mean	Standard Deviation
Age	Age of the respondent in years	58.90	12.12
Lbid	Natural log of the proposed bid value	8.57	0.57
Income	Annual income of the respondent's household (in ten thousand yen)	487.76	259.49
Sex	Gender of the respondent. Male =1, Female = 0	0.49	0.50
Visit	Whether the respondent went to visit public parks at least once in a month. Yes =1, No=0	0.40	0.49
Yes	Yes =1 if they are willing to pay and 0 otherwise	0.45	0.49

Table 6.6. Results of the model

Dependent Variable = WTP (probability of responding yes)

Variable	Model Coefficient (Wald Stat)
Constant	1.8838(0.4850)
Lbid	-0.6653(5.2664) [a]
Visit	-1.2906(14.0117) [a]
Income	0.0025(12.3353) [a]
Age	0.0546(11.8240) [a]
Sex	-0.1158(0.1133)
Initial log –likelihood value	270.058
Ending log –likelihood value	226.865
Model Chi-Square	43.194
Correct Predictions	71.94%
McFadden's R^2	0.16

Note: [a] Indicates that the coefficient is statistically significant at, at least, the .10 level

[b] McFadden's R^2 is one of the Psedudo-R-Square , varies between 0 and 1, is calculated as follows:

McFadden's R^2 = 1-[ending log-likelihood value/ initial log-like lihood value]

visit, income, and age are statistically significant. On the other hand, the sex coefficient is insignificant. According to the model chi-square statistic; the overall model is significant at the 0.01 level (critical value = 13.277 [df=3]) The model reasonably predicts 71.94% of the responses correctly. And also the McFadden's R^2 is calculated as 0.15.

6.6.2 Interpretation of Results

The regression coefficients of the Model with a log-linear valuation function can be interpreted loosely as the percentage change in the WTP for one-unit change in the level of each explanatory variable (Cameron 1988). Accordingly the willingness to pay for preserving public parks decreases dramatically if the bid value is increased and the respondent is a non-visitor (by about 67% and 129%, respectively). And for every yen rise in income, WTP is increased by about 0.25%. Also with an additional increase in years of age, WTP increases by about 5.46%.

Diagnostic tests were run on the multivariate models. The analysis of misclassification of individual observations showed only one misclassified cases providing an inadequate base for further analysis. As the number of misclassification is very low, no further diagnostic analysis is conducted (Hair et al. 1998).

6.7 Discussions

The results of the study lead us to the following observations and conclusions:
1. Both the average WTP of ¥5,225 and the aggregate WTP of ¥920 million signifies the overall importance of preserving public parks in Nagasaki City. This amount should be considered as a vital amount by the policy makers in re-thinking the need of developing and improving public parks in Nagasaki City.
2. As explained in Fig.6.4, approximately 50 percent of the respondents reported that, they do not visit public parks once in a month and absence of nature was one of the main reasons cited for not visiting. Also this was the third major reason for not willing to pay (see Fig.6.7).
3. The WTP model indicated that, willingness to pay is positively influenced by age and income and on the other hand negatively by bid value and visit. Thus, the model satisfies the trend of rational choice theory, as the WTP decreases with the increase in bid value and increases with increase in the ability to pay (income). Lastly, positive coefficient of the age variable signifies that social forces do play an important role in the willingness to pay.
4. Among these, the visit explanatory variable is very important in our study, which pointed out that the respondents' willingness to pay decreases if he/she is a visitor. This indicates that the non-visitors are willing to pay more for preserving the public parks, reflecting higher non-use value of public parks in Nagasaki City. Hence in addition to the use value (value from the current visitors), the non-use value (value from non-visitors) of public parks is also very high.

6.8 Public Park Case Study Conclusion

It can be concluded from the findings of this chapter that, environmental valuation techniques such as CVM used in this study might be considered as one of the alternatives to know the opinion of the residents in monetary figure and incorporate them in the do or not to do decision (benefit-cost analysis). Although public parks are not sufficient in the Nagasaki City compared to the national standard and average, but they are considered to be important both by present and potential users and are valued highly. Appropriate steps should be taken by policy makers to increase and preserve the public parks in the Nagasaki City with increased existence of natural environment. This would definitely make life in Nagasaki City more comfortable and raise its livability.

PART IV
CASE STUDY 3:
THE HUIS TEN BOSCH

CHAPTER 7 Overview of the HTB

7.1 Introduction

Huis Ten Bosch, meaning "House in the Forest" in Dutch, is one of the biggest theme parks in Kyushu Island, Japan. It is a famous private recreational theme park opened on March 1992 at an approximate cost of ¥300 billion, created by transforming 152 ha of industrial wasteland through various environmentally affable mechanisms (HTB 2002). The concept of Huis Ten Bosch is "coexistence between ecology and economy" and opened in accordance with a law to promote the construction of resorts with the support of the central and prefectural governments (see Fig. 7.1 for the location of the HTB).

Modeled after a Dutch town, the resort and theme park was once considered as popular as Tokyo Disneyland. Huis Ten Bosch attracted 4.25 million visitors in 1996 at the peak of its popularity, and 3.55 million in 2001 (see Fig. 7.2). The 152 ha park is twice as large as Tokyo Disneyland, and has about 1,000 employees (Yomiuri, 2003a).

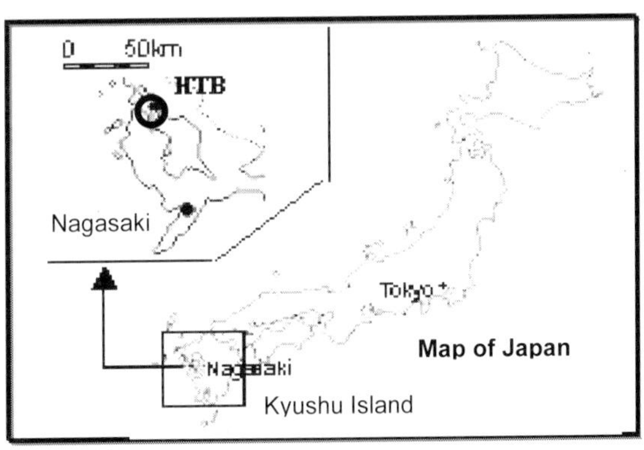

Fig. 7.1. Location of the HTB

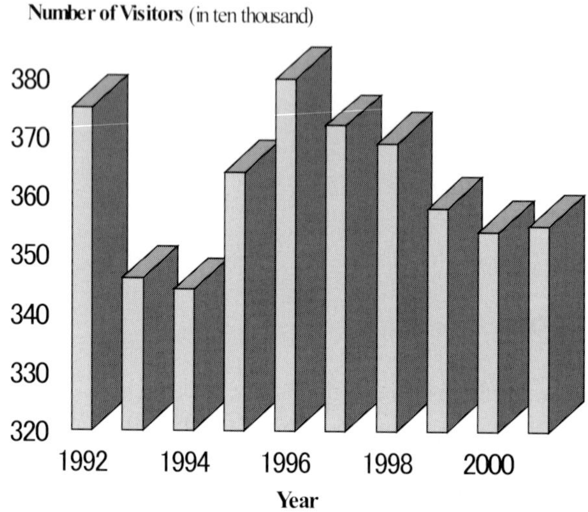

Fig.7.2. Yearly number of visitors visiting HTB
Source: Derived from Huis Ten Bosch Environmental Accounting Report: 1992-2001

7.2 Cost-benefit Analysis of the HTB

In the year 2002, HTB completed its 10th year of operation. Accordingly, they have published *Huis Ten Bosch Environmental Accounting Report 1992-2001*. In this report, environmental cost-benefit analysis of HTB during these years are discussed and calculated the total cost of the project on environmental purposes as about ¥49 billion (see Table 7.1). On the other hand, regarding the benefits they only calculated the benefits received by HTB itself through various environmental activities as about ¥23 billion however failed to quantify the social benefits provided by HTB (HTB 2002).

7.3 Current Status of the HTB

Although Huis Ten Bosch marked a bright start by attracting visitors not only form Japan, but also from neighboring Asian countries, this trend did not continued. HTB started to fall into financial difficulties for the following reasons (Yomiuri 2003b):

1. Fall in the number of visitors as they began to favor destinations other than Kyushu.

2. Intensified competition from rivals, including Universal Studios Japan, Osaka, and Tokyo Disney Sea.
3. Higher investment of ¥100 billion forcing to incur deficit every year.
4. Side business to sell plots of land for villas ended up in red, making the financial distress more severe.

Accordingly, HTB revenues continued shrinking and stood at about ¥35 billion in 2001, down 40 percent from the peak annual revenues of about ¥49.6 billion in 1996, while the liabilities stood at about ¥229 billion, including about ¥180 billion in loans (Yomiuri 2003a). Eventually on February 2003, operators filed for protection from its creditors under the Corporate Rehabilitation Law. The collapse of HTB is second in size in Japan, as a resort park to that of Phoenix Resort Ltd., the operator of Seagair Resort Park in Miyazaki Prefecture, which went under in February 2001. A list of the theme, amusement parks failed in Japan after 1998 is provided in Table 7.2. HTB is the 12th largest amusement park to go bankrupt in the past five years. Currently efforts are continuing for its revival under Corporate Rehabilitation Law.

Table 7.1. Environmental cost and benefits of Huis Ten Bosch (March 1992 ~ March 2002) (In million yen)

Cost			Benefit		
			Benefits to HTB		Benefits to the Society
Description	Fixed Cost	Accumulated Operating Cost	Description	Economic Benefits	
Restoring the ecosystem	5,068	1,887			Not Monetarily Quantified by the HTB Authority ?
Waste water treatment	8,399	1,266	Reduction in Electricity Energy cost	31	
			Reduction in water supply cost by recycling	916	
			Reduction in water cost by conversion of sea water to fresh water	127	
Clean Energy by co-generation system	6,881	7,463	Electric power generation	1,622	
Garbage separation and recycle	89	3,602	Reduction in Electricity cost required for garbage recycling	20	
Building Infrastructure	6,386	5,171	Increase in sales due to environmental friendly activities (32.4%)	19,575	
Automation	1,750	200			
Environmental study program	226	22	Income from Environmental study program	60	
Environmental community management	-	266	Free publicity from newspaper articles	183	
Total Costs	28,799	19,877	Total Benefits	22,534	

Source: Huis Ten Bosch Environmental Accounting Report: 1992-2001, HTB 2002

Table 7.2. Theme or amusement parks failed in Japan after 1998

Name	Location	Year of closing/ bankruptcy
Alpha Resort Tomamu	Hokkaido	May 1998
Sayama Yuen	Osaka	April 2000
Reoma World	Kagawa	September 2000
Gulliver's Kingdom	Yamanashi	October 2000
Enakyo Wonderland	Gifu	December 2000
Wild Blue Yokohama	Yokohama	February 2001
Yokohama Dreamland	Yokohama	February 2001
Phoenix Seagair	Miyazaki	February 2001
Tokyo Marine	Tokyo	September 2001
Namegawa Island	Chiba	September 2001
Mukogaoka Yuen	Kawasaki	April 2002
Huis Ten Bosch	Nagasaki	February 2003

Source: Compiled from 'Outlook grim for Dutch village theme park', Feb 28, 2003, The Daily Yomiuri, p.16.

CHAPTER 8 Environmental Valuation of the HTB

8.1 Introduction

Huis Ten Bosch (HTB) is in serious financial distress as evident from the discussion in Chapter 7. On this background, this chapter aims to estimate the environmental value of Huis Ten Bosch (HTB) by applying contingent valuation method with different elicitation methods. By applying this method in previous chapters we have estimated the value of the eco-system of the Isahaya Bay Wetland and conducted the CBA. Based on the experience of this Isahaya Bay Wetland study, in this chapter different elicitation methods are applied and conducted the HTB valuation study. This will help us to compare the variation in willingness-to-pay due to the change in the method of asking questions.

8.2 Justification for Using CVM

As we have discussed earlier in Chapter 2, CVM can be used to estimate economic values for almost all kinds of ecosystem and environmental amenities including recreational amenities. It possesses the rare advantage of estimating both the use and non-use (passive) values. In case of environmental goods like the HTB, non-use value is also important as it has potential high value for non-users too. Thus, it is appropriate to use CVM for estimating environmental value of HTB.

8.3 Survey Description

8.3.1 Study Region and Selection of Respondents

The questionnaires were sent to a random sample of 950 households of Sasebo and Nagasaki City. Mail survey technique was used for data collection and house-

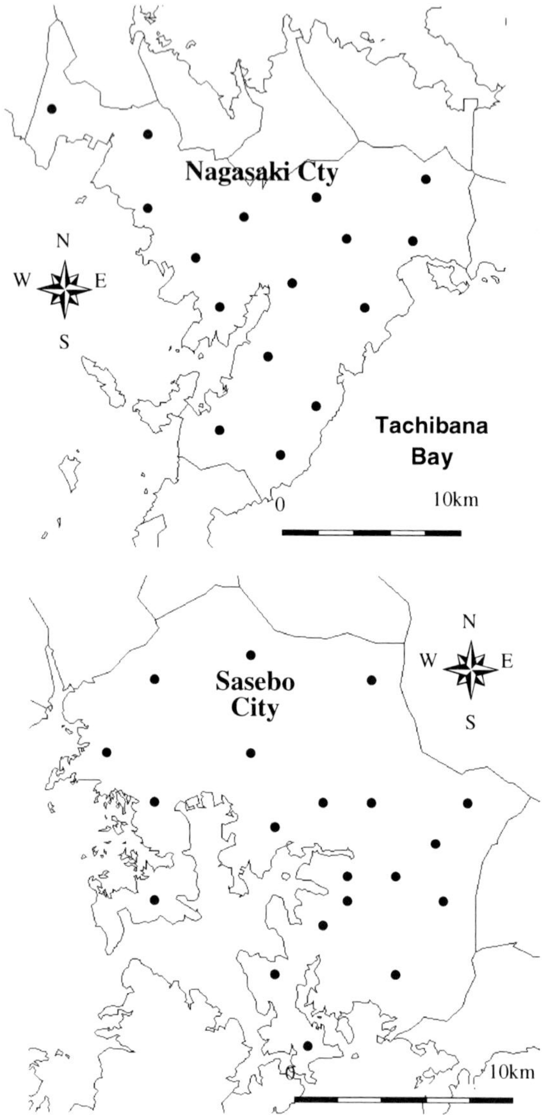

Fig. 8.1. Sampled areas in Nagasaki and Sasebo City (shown by black dots)

holds were selected randomly from the registered telephone directory (NTT, 2002). The sample cities and areas are shown in Fig. 8.1.

8.3.2 Response Rate

As shown in Table 8.1 the response rate is approximately 27 percent and 24 percent for Sasebo and Nagasaki City, respectively. It might be mentioned here that, getting higher response rate above 25 percent is difficult under mail survey technique of data collection in the perspective of respondents in Japan. But based on the experience of the mail survey response rate of studies conducted here, the response rate in our study can be considered to be satisfactory for conducting environmental valuation study (Ahmed et al. 2002b; Gotoh et al. 2002).

8.3.3 Contents of the Questionnaire

The original questionnaire contained different environmental valuation questions on Huis Ten Bosch and summarized in Table 8.2.

8.4 Findings

8.4.1 Visit to HTB

As shown in Figs. 8.2 and 8.3, the percentage of respondents visited HTB is very high, and among them a large portion also visited several times (80 percent and 48 percent in Sasebo and Nagasaki, respectively). On the other hand, the percentage of respondents did not visited HTB, at least once, was very low in Sasebo, only 6 percent and in Nagasaki this was 25 percent. Accordingly, it can be said that the majority of the respondents knows about the HTB, which is very important for CVM study.

8.4.2 Means of Transportation

The majority of the respondents from Sasebo and Nagasaki cities use cars for going to HTB (79 percent in Sasebo and 59 percent in Nagasaki). Very few use bus or train as means of transportation for going to HTB (see Figs. 8.4 and 8.5). This might be attributed to the closeness of HTB site from the respondents' residence.

Table 8.1. Response rate of the survey

	Sasebo City	Nagasaki City
Questionnaire Distributed	950	950
Filled Questionnaire Returned	252	232
Response Rate	27%	24%

Table 8.2. Contents of the questionnaire used in the HTB survey

Category of questions	Content of the questions
A Current status of public parks	A detailed description regarding the history, establishment and current status of Huis Ten Bosch (HTB) were provided, with the aid of pictures, graphs and supporting data.
B Warm-up questions regarding the image of public parks	1. Regarding visit to HTB? If yes, then how many times and by what means? 2. Knowledge of the relevant terms such as - the Drainage System, Desalination plant, Recycle Combust System and Biological Tropical System. 3. Image of the Huis Ten Bosch. 4. Influences from Huis Ten Bosch.
C WTP elicitation questions	5. Suppose we need to create a fund named 'Huis Ten Bosch Protection Fund' for salvaging HTB from financial distress by giving a value to the environmental activities initiated by it. Would you be willing to contribute ¥1,000 (assigned randomly ranging from ¥1,000 to ¥15,000) to this fund only once, in order to ensure sustainability of HTB? (This was followed by a follow-up question where the amount is increased or decreased, depending on whether the respondent's initial answer was positive or negative, respectively). 5-1. Reason for willing to pay 5-2. Reason for not willing to pay 6. Maximum Willingness to pay (WTP)
D Socio-demographic questions	7. Gender and family member 8. Age 9. Profession 10. Annual Income 11. Name of the town, period of stay, and distance from HTB 12. Free comments

Table 8.3. Bid values used under double-bounded dichotomous-choice elicitation method

Versions	Starting bid value (Yen)	Second bid value	
		YES (Yen)	NO (Yen)
First	1,000	3,000	500
Second	3,000	5,000	1,000
Third	5,000	7,000	3,000
Fourth	7,000	10,000	5,000
Fifth	10,000	15,000	7,000

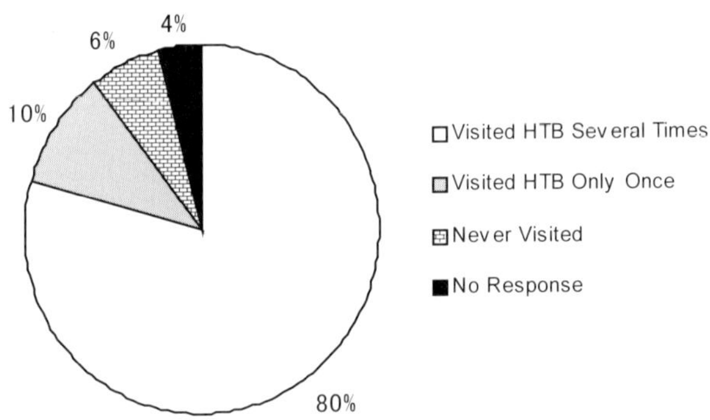

Fig. 8.2. Percentage of respondents visit to HTB for Sasebo City

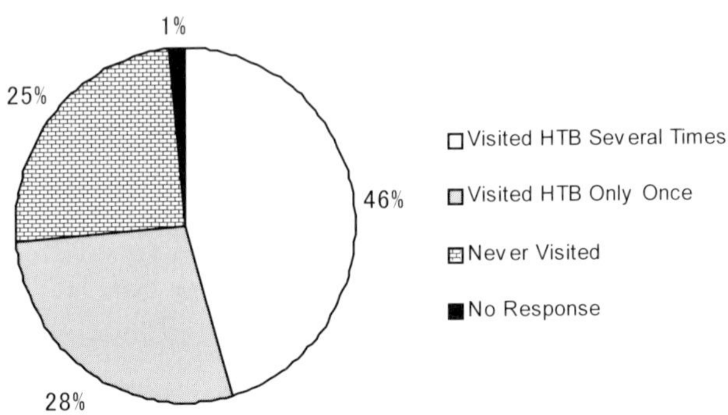

Fig. 8.3. Percentage of respondents visit to HTB for Nagasaki City

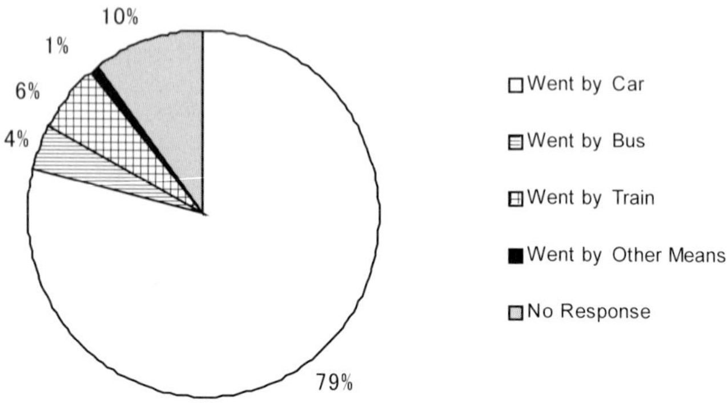

Fig. 8.4. Transportation means used by respondents for Sasebo City

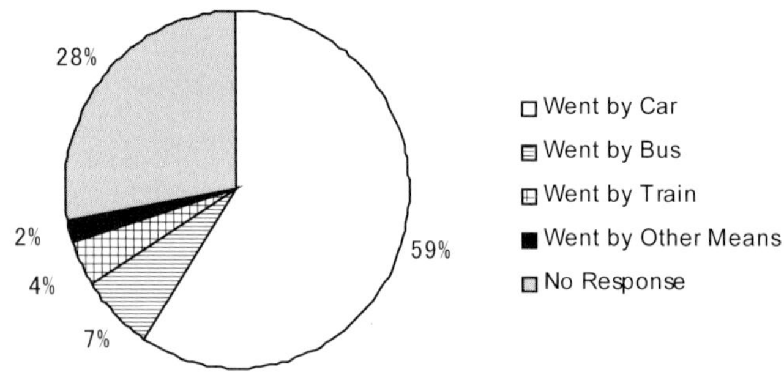

Fig. 8.5. Transportation means used by respondents for Nagasaki City

8.4.3 Substitute Site

The substitute site question made in the questionnaire revealed that Tokyo Disney Land (TDL) is the biggest rival of HTB. About 58 percent in Sasebo and 50 percent of the respondents in Nagasaki mentioned that, they view TDL as the substitute for HTB (see Fig. 8.6). The other favorite sites for them are Tokyo Disney Sea (TDS) and Universal Studio Japan (USJ).

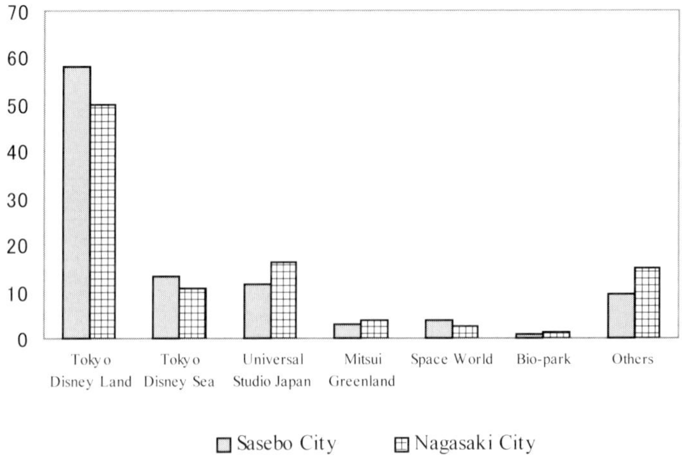

Fig. 8.6. Substitute site of HTB

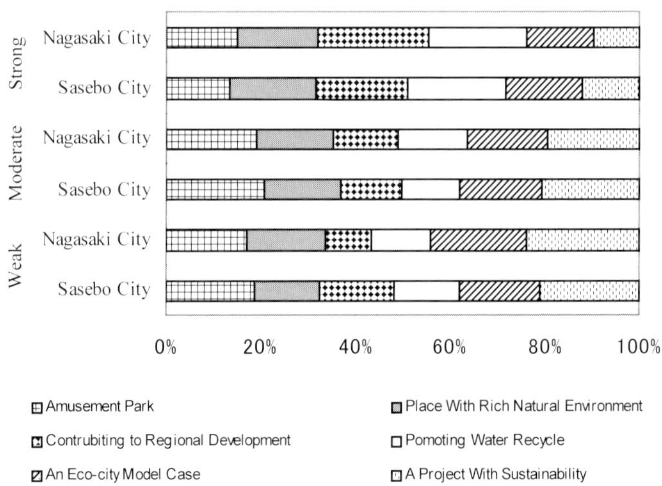

Fig. 8.7. Image of HTB

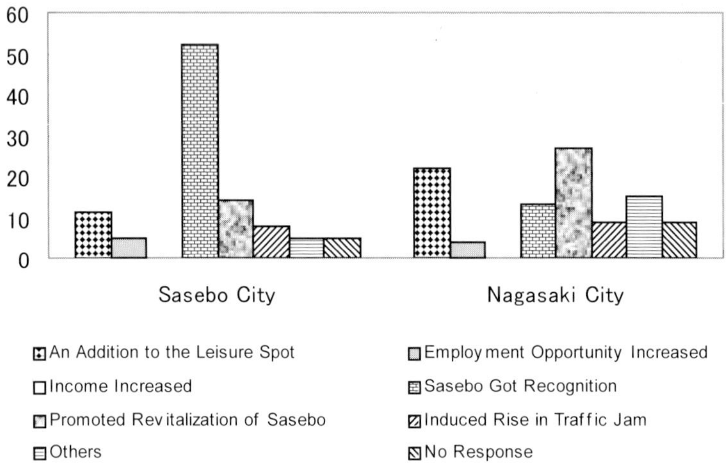

Fig. 8.8. Influence caused by HTB

8.4.4 Image of HTB

Regarding the image of HTB, according to the respondents of Sasebo and Nagasaki, the strongest images are: *contributing to regional development, promoting water recycle* and so on (see Fig. 8.7). On the other hand, the weakest images were reported to be a project with sustainability, an eco-city model case and so on, respectively. These findings indicate that, the respondents view HTB as an environmental friendly theme park contributing towards regional development, but the long-term sustainability of which is doubtful.

8.4.5 Influence Caused by HTB

While answering to the question regarding the influence caused by the HTB, positive influence was high (see Fig. 8.8). For example, 52 percent of the respondents answered that HTB earned Sasebo City recognition nationally and internationally, and 27 percent of the respondents of Nagasaki City responded that, HTB is serving in revitalizing Sasebo. While regarding negative impact, 8 percent and 9 percent of the respondents in Sasebo and Nagasaki, respectively, reported that HTB increased traffic jam in the city. No one in both Sasebo and Nagasaki responded that HTB increased their income.

Table 8.4. Willingness to pay

	Sasebo City	Nagasaki City
Sample Size (complete)	251	228
SD of the Mean	346	321
Mean WTP (¥/household)	4,606	3,523
Median WTP (¥/household)	2,000	2,000
Range of 95% confidence interval (¥)	251	228

Table 8.5. Total estimated WTP

	Sasebo City	Nagasaki City
Mean WTP (¥/household)	4,600	3,500
Total Number of Households (in thousand)	90	168
Estimated Total WTP (in million yen)[1]	414	588

[1]Estimated Total WTP is calculated by multiplying Mean (WTP) per household by number of household in the respective cities.

8.4.6 Willingness to Pay

The analysis of data is conducted by applying *Turnbull method*. The respondents who had resisted to pay in all of the two bids are considered to have zero WTP, but their opinion regarding refusal to pay are summarized in order to have the reasons for not willing to pay. Table 8.4 summarizes the results under Double bounded dichotomous choice (DBDC) elicitation method. From the Table, we can see that, the mean WTP is approximately ¥4,600 and ¥3,500 in Sasebo and Nagasaki, respectively, while the median WTP is ¥2,000 in both the cities.

8.4.7 Total Estimated WTP

Finally, if these figures are multiplied by the number of households in each respective city, then we can find that, HTB is worth approximately ¥414 million, and ¥588 million to the residents of Sasebo and Nagasaki City, respectively (see Table 8.5). If we compare the results of the study with other valuation study conducted by using CVM, then it can be concluded that the mean WTP estimated for both the respondents of Sasebo and Nagasaki is relatively lower.

8.4.8 Reasons for Willing to Pay

Figure 8.9 summarizes the reasons why the respondents are willing to pay for preserving the HTB. From the figure we can see that, need for protecting HTB for regional development is the most common reason quoted by the respondents, as the reason for payment (75 percent and 67 percent in Sasebo and Nagasaki, respectively). The need for future environmental preservation and attracting tourists to Sasebo are the other major reasons cited by the respondents for willingness to pay.

8.4.9 Reasons for Not Willing to Pay

Among the reasons for not willing to pay, 30 percent and 36 percent of the respondents of Sasebo and Nagasaki City expressed their unwillingness to pay as a fund (see Fig. 8.10). On the other hand, unlike the case study of IBW (see Chapter 4), very few respondents cited that the proposed amount is high as the reason for not willing to pay (12 percent and 13 percent in Sasebo and Nagasaki, respectively). This indicates that the set bid values for payment were not beyond their reach for payment; rather they did not felt interested to pay for other reasons.

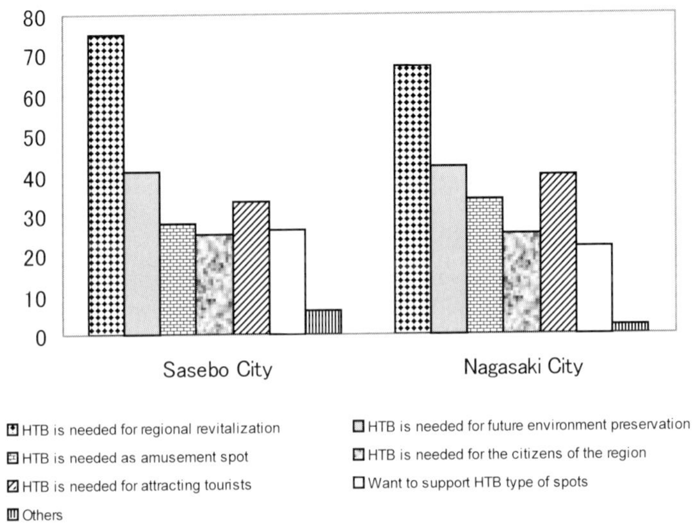

Fig. 8.9. Reasons for WTP

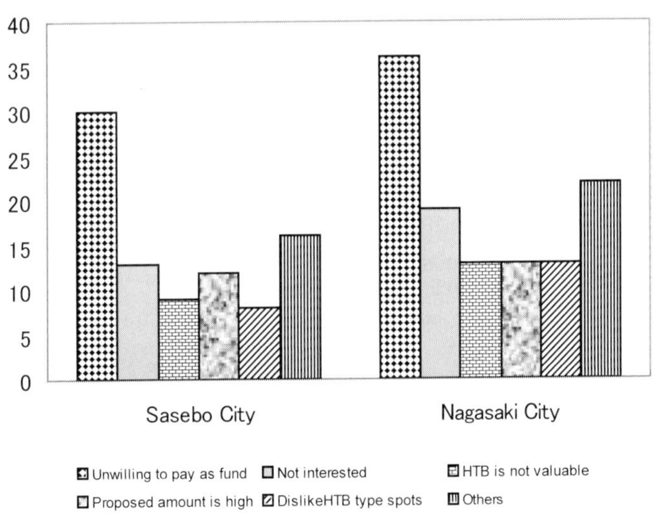

Fig. 8.10. Reasons for not WTP

8.5 Summary and Conclusion

8.5.1 The Reason for Lower WTP

The average WTP estimated for HTB might be considered as lower as similar CVM survey conducted on estimating the WTP for protecting public parks in Nagasaki City was ¥6,500 (Gotoh et al. 2003). The reason for this can be thought as following:

1. **Private theme park:** As also evident form the free comments, a major portion of the respondents viewed HTB as private theme park and expressed the view that, if it has any difficulties it should survive by its own ability.
2. **Identity problem**: The image of a theme park is strong for HTB among the respondents and this is hindering people to evaluate it as a environmental goods. Thus while valuing HTB as an environmental amenity, the respondents are prone to place a lower price.

8.5.2 Difference in WTP Between Nagasaki and Sasebo City

WTP in Sasebo city is comparatively higher than in Nagasaki City. One of the reasons for this, might be the view of the respondents of Sasebo regarding HTB as something that is linking towards the revitalization of Sasebo and attracting tourists from different places.

8.5.3 Recommendations

More active publicities are needed to introduce the environmentally friendly initiatives taken by HTB. As our survey study revealed that even the residents of Sasebo City are not so much familiar about HTB. As a large portion of the respondents felt that the entrance fee, attraction fee, hotel rent and foods are expensive inside HTB, necessary actions should be taken for making HTB more cost effective. The following are the suggested measures:

1. In order to raise the number of repeaters, the entrance fee for getting inside HTB should be kept as minimum as possible.
2. There should be special discount group prices for the pupils or students of school and university, and also for general public for supporting environmental education.

CHAPTER 9 Calculating Cost-Benefit Analysis of the HTB

9.1 Introduction

On June 2002, Huis Ten Bosch (HTB) authority had published the *Huis Ten Bosch Environmental Accounting Report: 1992-2001*. Here they calculated the *total cost* of the HTB project for environment friendly amenities until March 2002 (see Table 9.1 for details). Table 9.1 summarizes the main headings of this environmental cost-benefit analysis. We can see from this estimation that the total environmental costs incurred since the inception of HTB is estimated about ¥49 billion yen including the establishment and operating costs. Against these environmental costs, HTB received a total of about ¥23 billion benefits. But as HTB authority claims and we also agree with them that, these costs for environmental purpose not only brought benefits to HTB, but also to the society. Now the question is how to measure these benefits.

As a solution to this problem, in this chapter a modest attempt is taken to estimate the benefits received by the society from HTB by applying contingent valuation method (CVM) and aimed to fill out the question marks shown in Table 9.1. Hence, the main objective of this chapter is to conduct the cost-benefit analysis (CBA) of HTB by incorporating environmental valuation considerations to judge its future sustainability if the economic performance get worsen.

9.2 Analytical Methodology

The same logic, steps and methodology are applied by using CVM in the cost-benefit analysis (CBA) of HTB (see Chapter 4 for details). Also in estimating the mean WTP, the Turnbull method of non-parametric estimation is followed as described in the previous chapters (see Chapters 4 and 5 for detailed methodology). The database used in this section is also the same as that of Chapter 8, with the only difference that, here the data of Sasebo and Nagasaki are accumulated under one set and estimated the aggregate WTP for the two cities.

Table 9.1. Recalculation of cost-benefit Analysis of the HTB

Category	Amount (Million Yen) / Ratio
1. Total environmental costs	48,676
2. Total environmental benefits to HTB	22,534
3. Total environmental benefits to Society	?
4. Total environmental benefits	?
5. Benefit-cost (BC) ratio	?

Table 9.2. Total Estimated WTP

Description	Number/ Amount
Sample size (complete)	479
Mean WTP (¥/household)	4,088
Median WTP (¥/household)	2,000
SD of the Mean	238
Range of 95% confidence interval (¥)	±467
Total Number of Households in Kyushu	4,976,000
Estimated Total WTP (Billion yen)	20.34

9.3 Deriving Total Environmental Benefits of HTB

On the basis of the methodology described above, the mean (WTP) is estimated to be approximately ¥4,088 per household for all the sample respondents of Sasebo and Nagasaki City. (See Table 9.2). The median (WTP) is ¥2,000 per household.

Figure 9.1 demonstrates the percentage of the respondents' willing to pay at various amounts for preserving Huis Ten Bosch. As evident from the negatively sloped curve that the percentage willing to pay decreases with the increase in amounts to pay.

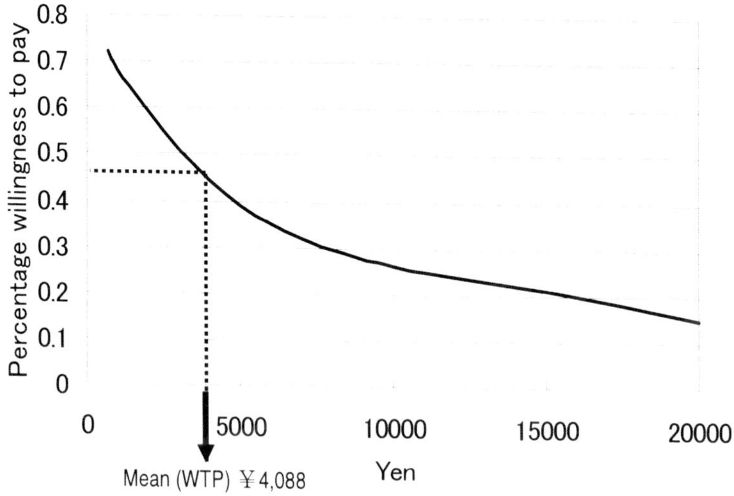

Fig. 9.1. Percentage of respondents WTP

Table 9.3. Recalculation of cost-benefit analysis of the HTB

Category	Amount (Million Yen) / Ratio
1. Total environmental costs	48,676
2. Total environmental benefits to HTB	22,534
3. Total environmental benefits to Society	20,340
4. Total environmental benefits	42,874
5. Benefit-cost (BC) ratio	0.88

9.4 Calculating CBA

Next we are going to extrapolate the mean WTP of ¥4,088 to the entire 4,976,000 (MPHP 2003) private households of Kyushu to estimate the benefits received from various environmental costs incurred by HTB. Accordingly, it was found that, the residents of Kyushu are willing to pay approximately ¥20 billion to preserve the Huis Ten Bosch (See Table 9.2).

In addition, the benefit-cost ratio of the HTB is calculated in Table 9.3. This table is the continuation of Table 9.1, with the same cost and benefit amounts. The only difference is that, here ¥20 billion calculated above is inserted as the socio-environmental benefit of HTB. Inclusion of such benefits increases the total benefits from the environmental investments of the project to about ¥43 billion (See Table 9.3). As a result of this, B/C ratio of ten years operation of HTB becomes 0.88.

9.5 Conclusion

In this chapter, a very modest attempt is taken to suggest ways to incorporate the benefits derived by the society into the cost-benefit analysis of Huis Ten Bosch by applying CVM. Inclusion of social benefits of ¥20 billion raises the total environmental benefits of HTB to about ¥43 billion. This reduces the gap between total environmental costs and benefits to ¥5,802 million. Thus it can be said that it would take only another 2.8 years to cover up the fixed cost incurred for various environmental amenities by yearly environmental monetary benefits received by HTB from these facilities.

9.6 HTB Case Study Conclusions

Through the results of the environmental valuation and cost-benefit analysis of HTB conducted in Chapters 7,8 and 9, it can be concluded that, HTB has higher value and thus chance of sustainability as an environmental conservation park rather than a theme park. Because to the respondents of our study, HTB has a social benefit value of about ¥20 billion not as a theme park, but as environmental conservation spot. More publicity and coordinated efforts are needed to influence the people to recognize HTB as not only a theme park, but also an environmental amenity, which are required to be protected. HTB should make its identity clear to the public as whether it's a theme park or an environmental amenity. Because, if it is a theme park then it should survive by its own efforts and must face the competition usually prevalent in a market economy. On the other hand, if it can validate its identity as an environmental amenity, then it can gain support from the general people for ensuring its existence, and thus, have different valuation considerations.

PART V:
NEW APPROACHES FOR ENVIRONMENTAL VALUATION

CHAPTER 10 Effect of Distance on Willingness to Pay

10.1 Introduction

The relationship between geographical distance and the willingness to pay for preservation and improvement of particular environmental goods are generally thought to be negative. As shown by the negatively sloped curve in Fig. 10.1, various studies including Sutherland and Walsh (1985), Hannon (1994), Pate and Loomis (1997) also have found this relationship to be negative and argued through empirical analysis. According to them the more away the respondent resides form the area, the less likely he/she would be willing to pay for improvements or conservation of it. Although these arguments are logical, it is difficult to accept this relationship to be universal. Because environmental goods in different countries have different surrounding environment and various judgment work behind their preservation. Our first case study 'the Isahaya Bay Wetland' is famous in Japan for its difficult and complex nature of the problem. Hence, it is interesting to verify whether negative relationship also exists between distance of the respondent to the area and the willingness to pay for preservation of it.

10.2 Data and Methods

The database of our first case study i.e., the Isahaya Bay Wetland study is used to construct a multivariate model supported by economic theory. In doing so, different variables are gradually added, in addition to our main concern variable distance, to identify those having influence on WTP, by constructing three different models. Finally, one model is selected because it has the highest level of predictability power and the significant variables are isolated and interpreted. Subsequently, the results of the multivariate analysis is verified by running diagnostic tests for identifying the existence of outliers, if any, and ensuring the robustness of the full model.

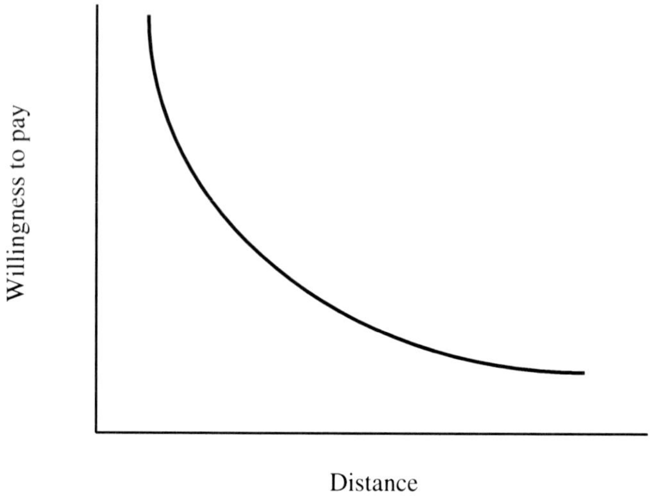

Fig. 10.1. Conventional relationship between distance and WTP

10.3 Multivariate Analysis

10.3.1 The Model

A multivariate model is constructed to determine the factors, which influences willingness to pay for restoring the Isahaya Bay Wetland. It comprises the variables that are assumed to have an influence on willingness to pay, such as: bid value, visit, income, distance and demographics[1]. Accordingly, bid value and income are included to test the "*rational choice theory*". Visit is included to test the impact of *non-use value*. Distance is included to see its impact on WTP. Finally, demographic variables such as age and sex are included to check whether *social forces* play a role in the willingness to pay. These factors, which are hypothesized to have influence on WTP based on economic theory, can be summarized by the following conceptual model:

$$WTP = f(\text{bid, visit, income, distance, demographics}) \quad (10.1)$$

In case of CV studies with dichotomous structure of the dependent variable, the model can be estimated through a non-linear probability model. The most commonly used is the logit model (Pate and Loomis 1997). Because of the discrete na-

---

[1] For model selection and logistic regression analysis, see Raftery 1995; Rice 1994: Stynes 1984; Tabachnick 1996

ture of the dependent variable, the ordinary least square regression can be used to fit a linear probability model, but its heteroskedastic nature of the error terms and the possibility to predict values beyond the range of 0 and 1, led us to use the logistic regression model. Accordingly, the logistic regression model, which can be developed to analyze hypothesized model in Equation 10.1 takes the following shape:

$$\text{Ln}[\text{prob(yes)}/\{1\text{-prob(yes)}\}] = b_o - b_1 \text{ (bid)} + b_2 \text{ (visit)} + b_3 \text{ (income)} + b_4\text{(Ldistance)} + b_5\text{(age)} + b_6\text{(sex)}) + b_7\text{(famem)} \quad (10.2)$$

The explanation of the variables and the descriptive statistics are presented in Table 10.1, with their expected relationship sign with the willingness to pay.

10.3.2 Descriptive Statistics and Model Selection

The results of the descriptive statistics are presented in Table10.1 and logistic regression results are shown in Table10. 2. Three models are constructed by including different blocks of independent variables, to examine the effects of respective variable on the overall predictability of the model.

Table 10.1. Descriptive Statistics (n =390)

Variable name	Mean	Standard deviation	Description	Expected sign in regression model
Yes	0.38	0.48	The dependent variable is the probability of yes response to bid amount (Yes=1 if they are willing to pay and 0 otherwise).	
Age	22.27	22.40	Age of the respondent in years	?
Bid	7197.44	4557.49	The bid amount first offered to the respondent.	–
LDistance	66.13	63.66	Natural log of distance between respondent's home and the IBRP	?
Famem	2.95	1.36	The number of family member of the respondent's household.	–
Income	4.66	2.79	Annual income of the respondent's household divided by 10 million and considered as covariates as it as 14 categories	?
Sex	0.48	0.50	Dummy variable representing respondent's gender	?
Visit	0.45	0.50	Dummy variable representing whether the respondent went to visit the IBW at least once	?

Table 10.2. Results of the three models

Dependent Variable = WTP (probability of responding yes)

Variable	Model 1 Coefficient (Wald Stat [a])	Model 2 Coefficient (Wald Stat [a])	Model 3 Coefficient (Wald Stat [a])
Constant	0.4114	-0.2082	0.0084
	(3.1520)	(0.5017)	(0.0004)
Bid	-0.001	-0.001	-0.0001
	(17.9601) [b]	(19.5963) [b]	(19.3192) [b]
Visit	-0.2613	-0.5775	-0.4648
	(1.4865)	(5.9080) [b]	(3.3929) [b]
LDistance		0.2450	0.2819
		(12.0831) [b]	(8.5598) [b]
Income			0.1897
			(17.8625) [b]
Age			-0.0117
			(2.7546) [b]
Sex			-0.3045
			(1.7181)
Famem			-0.2959
			(10.5218) [b]
Initial log – likelihood value	520.626	520.626	520.626
Ending log – likelihood value	500.108	487.313	460.847
Model Chi-Square	20.518	33.314	59.779
Correct Predictions	64.10%	65.87%	68.72%
McFadden's R^2	0.039[c]	0.064[c]	0.115[c]

Note: [a] The Wald stat is simply the square of the t-statistic

[b] Indicates that the coefficient is statistically significant at, at least, the .10 level

[c] McFadden's R^2 is one of the Psedudo-R-Square , varies between 0 and 1, is calculated as follows:

McFadden's R^2 = 1-[ending log-likelihood value/ initial log-likelihood value]

In Model 1, only two variables are included: BID and VISIT. The results of the model indicate that, the respondent's willingness to pay decreases as the bid value increases. The coefficients on the BID have a Wald statistic equal to 17.96 and are significant at the 0.01 level (99% confidence level) with a critical value of 6.635 with 1 degree of freedom. On the other hand, the VISIT coefficient is insignificant. According to the model chi-square statistic, the overall model is significant at the 0.01 level. The model reasonably predicts 64.10% of the responses correctly. And also the McFadden's R^2 is calculated as 0.039.

Model 2 includes one more additional variable: LDISTANCE. Now the coefficient of BID, VISIT and LDISTACE are statistically significant. The model chi-square statistic is also significant at the 0.01 level. The VISIT variable becomes statistically significant in this model. The percent of corrected predictions increases by1.77%, and the McFadden's R^2 is approximately 64% higher. All these showed the overall superiority of Model 2 to Model1 in terms of model fit and it can be said that the inclusion of the additional variable LDISTANCE enhanced the predictability of willingness to pay. Finally in Model 3, income and the demographic variables such as: AGE, SEX and FAMEM are incorporated. The model chi-square implies that the Model 3 is comparatively superior to Model 2. According to the Wald stat INCOME, AGE and FAMEM are significant, while, SEX is not significant. According to the model chi-square statistic; the overall model is significant at the 0.01 level (critical value = 16.812 [df=6]) The percent of correct prediction also increases by almost 3% and McFadden's R^2 statistic increases by about 80%. All these statistical indicators suggest that the inclusion of demographic variables improved the model and as such Model 3 is better than Model 2, the one selected finally in this study.

10.3.3 Case-wise Diagnostics

Diagnostic tests were run on the multivariate models. The analysis of misclassification of individual observations showed only four misclassified cases providing an inadequate base for further analysis. As the number of misclassification is very low, no further diagnostic analysis was conducted (Hair et al. 1998).

10.4 Interpretation of the Model Results

The regression coefficients of our selected Model 3 with a log-linear valuation function can be interpreted very loosely as the percentage change in the WTP for one-unit change in the level of each explanatory variables (Cameron 1988). Thus, for every yen rise in INCOME, WTP is increased by about 18.97%. Also the willingness to pay for restoring IBW increases dramatically with the distance of the respondent's location from the IBRP by about 28.19%. This indicates that those who are residing near to the IBW are willing to pay less than those who are not, for restoring the eco-system of IBW. Conversely, the respondents living away form the IBW are less likely willing to pay for its preservation. Interestingly, Sutherland and Walsh (1985), Hannon (1994), Pate and Loomis (1997), as we have discussed in the introductory part of this chapter, found negative relationship between distance and willingness to pay. Finally, WTP decreases by about 46.48%, 1.17% and 29.59%, if the respondent visited IBW, additional increase in age, and addition to one member in the family, respectively.

10.5 Summary and Conclusion

The results of this chapter might be summarized and concluded as follows:

1) Inclusion of the explanatory variables such as: bid, visit, income, distance and demographics, increased the predictability power of the model. The multivariate analysis indicated that, willingness to pay is positively influ-

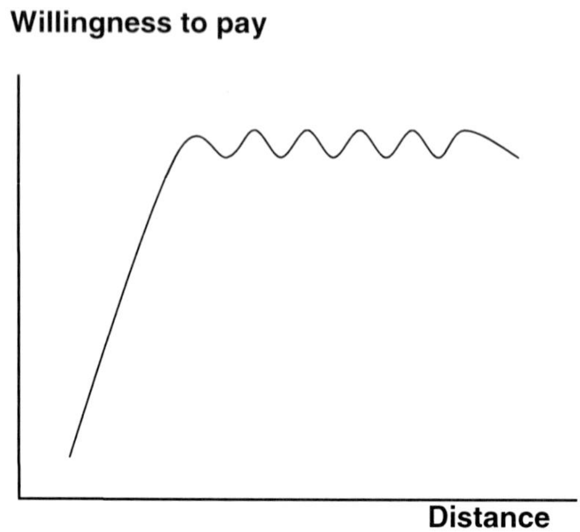

Fig.10.2. Predicted relationship between distance and WTP

enced by distance and income factors and on the other hand, negatively by bid value offered, visit, age and number of family members.

2) The result of the explanatory variable distance showed that the respondents' willingness to pay decreases as the distance from the IBW increases. This is a very important result obtained from this study, which provide results contrary to the conventional relationship between distance and willingness to pay in wetland valuation. The relationship can be explained by Fig. 10.2.
3) This positive relationship was also observed between distance and willingness to pay in Chapter 4 by calculating area wise mean WTP and found that both the mean and median WTP rises as we go far from the location of the Isahaya reclamation project.
4) Again in this section of the study this negative relationship is confirmed by multivariate analysis. Theoretically this can be explained by the fact that, typhoon, flood and high tide in various occasions caused heavy damages to life and property in the Isahaya Bay region namely, High Tide and Flood in 1914 and 1927, Isahaya Big Flood in 1957, Flood in 1982 and High Tide in 1985 (MAFF 1990). As a result, the residents living nearer to the project site support the IBRP, which is aimed at constructing sea dikes for protection from flood disaster in IBW. They are afraid of future devastation caused by natural calamities and thus do not want to pay to save the eco-system of the IBW, which would call for discontinuing the reclamation project. Hence, the residents living more closer to the IBRP site are likely to be more afraid about the flood and high tide disaster and are less likely be willing to pay for restoring IBW, by calling off IBRP. On the other hand, as we go far from the location of the Isahaya Bay, people are less affected by the flood disaster

and are willing to pay more to save the IBW as they only view the positive impact of preserving the eco-system and do not face the natural calamity challenges.

Therefore it can be concluded that, it is not always correct to say that, the further the respondent resides form the area, the less likely he/she would be willing to pay for the improvements or preservation of it. As we have seen, in case of the particular environmental good like the Isahaya Bay Wetland, where sea dikes are constructed having both positive (flood control) and negative (environmental damage) impact, the further we go from the location of the project, more the willingness to pay increases. However, this is merely the result of one case study. More detailed studies are needed to confirm the result of this chapter.

CHAPTER 11 Free Comments in CV Survey and Their Impact on WTP

11.1 Introduction

Almost in every contingent valuation method study, it is customary to include some space at the end of the questionnaire, in order to enable respondents to write their free opinion or comments regarding the contents of the survey or anything that they feel free to express. It is the only in this part, where respondents get the opportunity of free expression without any direction from the researcher and thus bears higher significance. But unfortunately, there are very few research papers, which brought these free comments into light by publishing them. This chapter of this book aims to explore this aspect by summarizing the free comments of environmental valuation survey conducted on the Isahaya Bay Wetland (details of the original study are provided in Gotoh et al. 2002 and Ahmed et al. 2002c, 2003).

In doing so, the comments of the respondents are presented in a graphical form and then attempted to verify the relationship between the free comments and the willingness to pay (WTP). The results of the study would bring the raw opinion of the respondents into glow and provide us more insight into this highly sensitive Isahaya Bay Wetland issue in Japan.

11.2 Data and Methods

11.2.1 Study Area and Response Rate of the Data Used

The survey data on environmental valuation of the Isahaya Bay Wetland is used, where a total of 1,800 questionnaires were distributed in the three cities of Kyushu Island- Isahaya, Nagasaki and Kitakyushu (600 each). The locations of the sample cities are shown in Fig. 11.1. Mail survey technique is used for data collection with households selected randomly from the telephone directory for each city (see Chapter 3 for survey description).

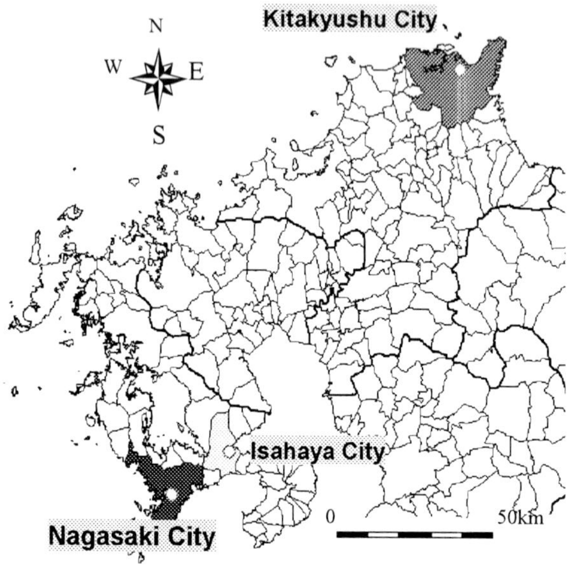

Fig. 11.1. Sample city covered

11.2.2 Analytical Tools Used

Two-step analysis is conducted in this study. Firstly, raw comments of the respondents are classified into five board headings and discussed accordingly. Secondly, from these free comments we sorted out whether the respondent are supporting the IBRP or otherwise and tried to review its relationship with WTP along with other relevant variables.

The best-subset approach of model selection is used for identifying the most influencing variable (Levine et al. 2002). Under this approach the procedure followed is outlined in Fig. 11.2. Accordingly, two criteria are used to choose among models with different combination of independent variables. The first criterion is variance inflationary factor (VIF), used for measuring collinearity among the explanatory variables. VIF_j, the variance inflationary factor for variable j is defined as follows.

$$VIF_j = \frac{1}{1-r_j^2}. \tag{11.1}$$

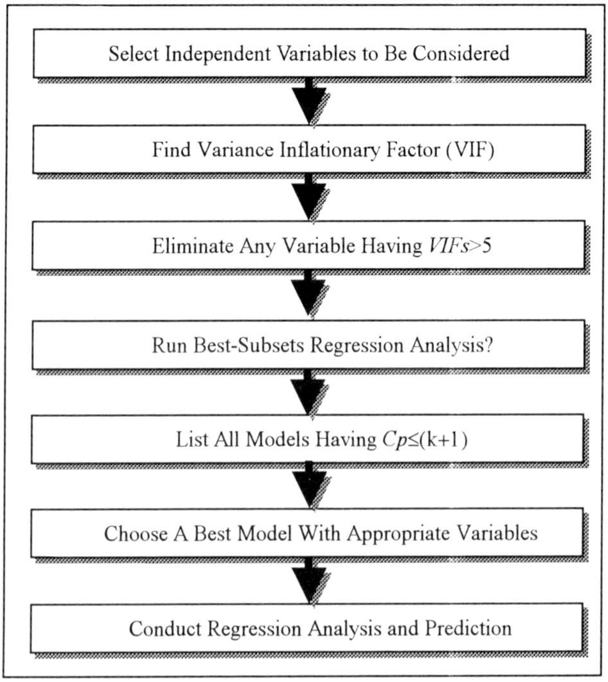

Fig. 11.2. Step-wise procedure in model building by best-subsets approach

Where, r_j^2 is the coefficient of multiple determination of explanatory variable X_j with all other X variables. As a rule of thumb all VIF should be less than 5.0 to prove that there is no collinearity among the explanatory variables (Snee 1973). A second criterion used for evaluating competing models is C_p statistic, developed by Mallows and is defined as shown in Equation 11.2 (Neter et al. 1996).

$$C_p = \frac{(1-r_k^2)(n-T)}{1-r_T^2} - (n-2(k+1)). \tag{11.2}$$

Where,
k = number of independent variables included in a regression model
T= total number of parameters (including the intercept) to be estimated in the full regression model.
r_k^2 = coefficient of multiple determination for a regression model that has k independent variables
r_T^2 = coefficient of multiple determination for a full regression model that contains all T estimated parameters.

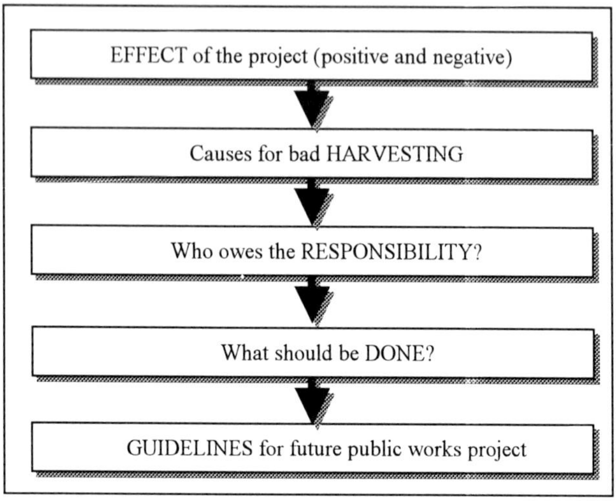

Fig. 11.3. Broad classification of free comments

As a selection standard the goal is to find models whose C_p is close to or below (k+1).

11.3 Summary of the Free Comments

The number of respondents commented are 70, 58 and 72 in Isahaya, Nagasaki and Kitakyushu, respectively (see Table 11.1). Free comments of the respondents are summarized in an easily understandable way and classified them into five broad headings as shown in Fig. 11.3 in the order of: effects, causes, responsibility, solution and guidelines, respectively. Then the comments classified in such a manner are presented through Figs. 11.4 to 11.9 and are explained below.

11.3.1 Effect of the Project (positive and negative)

Both positive and negative effects of IBRP are summarized in Figs. 11.4 and 11.5. The most frequently quoted positive effect of IBRP was protection from flood disaster (50 percent, 23 percent and 8 percent in Isahaya, Nagasaki and Kitakyushu, respectively. Although 30 percent of the respondents commented from Isahaya, viewed IBRP as protecting them from water logging during rainfall; none from Nagasaki or Isahaya viewed in the same way.

Among the negative effects of IBRP, destruction of eco-system for all the cities was most commonly mentioned comments (see Fig. 11.5).

Table 11.1. Number of respondents commented

	Isahaya City	Nagasaki City	Kitakyushu City
Total respondents	124	117	160
Number of respondents Commented	70	58	72

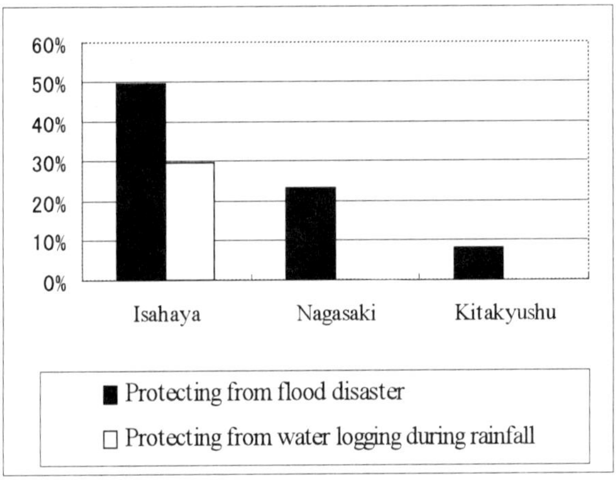

Fig. 11.4. Positive effects of IBRP

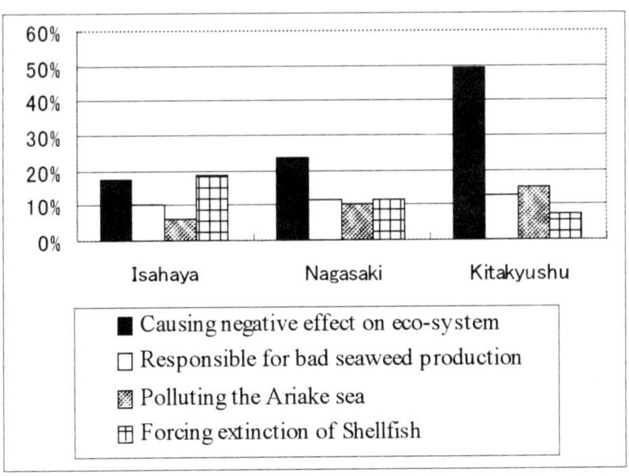

Fig. 11.5. Negative effects of IBRP

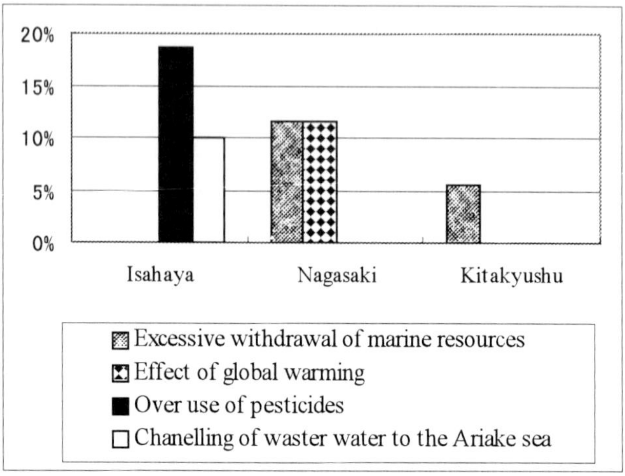

Fig.11.6. Causes for bad harvesting

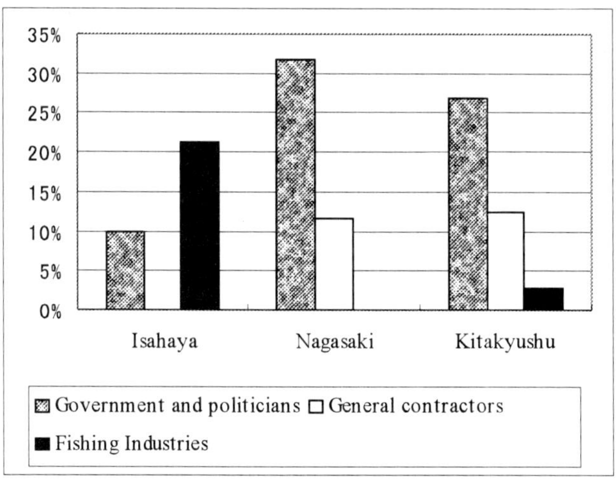

Fig.11.7. Who owes the responsibility?

11.3.2 Causes for Bad Harvesting

Regarding the causes of bad harvest of fisheries product, which is most frequently referred as the negative impact of IBRP, 19 percent and 10 percent of the respondents in Isahaya mentioned that, excessive use of pesticides and channeling of waste water to the Ariake sea, respectively (see Fig. 11.6). On the other hand, 12 percent and 6 percent of the respondents in Nagasaki and Kitakyushu,

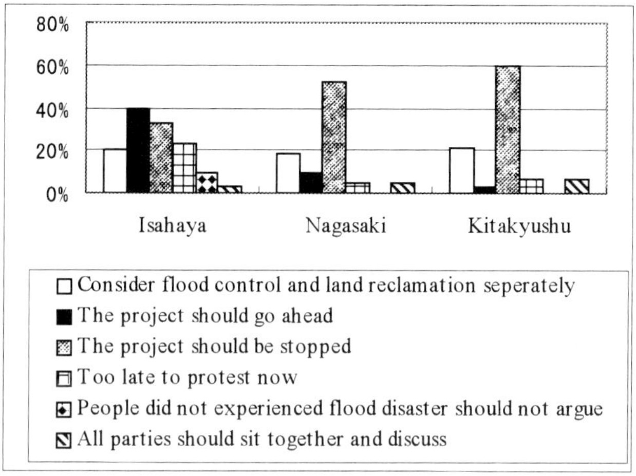

Fig.11.8. What should be done?

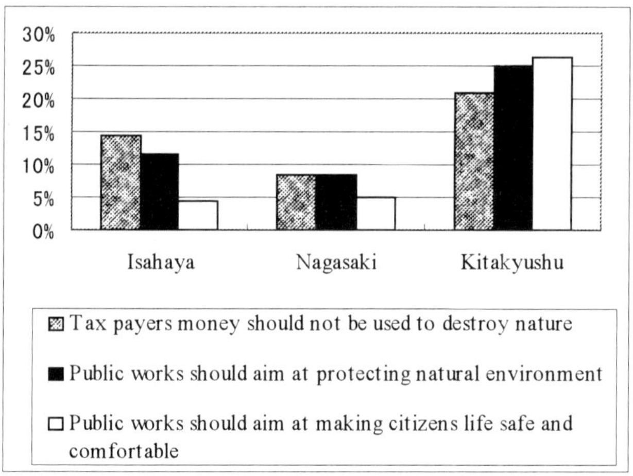

Fig.11.9. Guidelines for future public works project

respectively, commented that excessive withdrawals of marine resources are the reasons for the bad harvest.

11.3.3 Who Owes the Responsibility?

As shown in Fig. 11.7, respondents of all the cities, more or less, commented that the government and the politicians are responsible among others, for the crisis created by IBRP (10 percent, 32 percent and 27 percent in Isahaya, Nagasaki and Kitakyushu, respectively). On the other hand, 21 percent of the respondents commented from the Isahaya City that the fishing industries are also responsible for eco-system catastrophe in the region.

11.3.4 What Should Be Done?

Regarding the solution to this problem, the respondents are largely divided (see Fig. 11.8). A significant percentage of respondents of all the cities agreed that, the project should be stopped (33 percent, 52 percent and 60 percent in Isahaya, Nagasaki and Kitakyushu, respectively). On the other hand, as high as 40 percent respondents in Isahaya, 10 percent in Nagasaki and 3 percent in Kitakyushu commented that the project should be continued.

11.3.5 Guidelines for Future Public Works Project

Finally, in suggesting guidelines for public works project, respondents from all the cities commented that, public works project should aim at making citizens life safe, comfortable and at the same time protect the natural environment. They cautioned policy makers to make efficient use of their taxes (see Fig. 11.9).

11.4 Multivariate Model

To test the relationship between the free comments and the willingness to pay (WTP) we have sorted out different qualitative information into quantitative form and succeeded to trace out variables such as, *support*, *influence*, *visit*, and *years residing*. Then different demographic variables are add to construct alternative models and identify the variables which influencing the WTP most. Accordingly the conceptual model stood as follows:

$$\text{WTP} = f\ (\text{support, influence, visit, years residing, income, demographics}) \tag{11.3}$$

Where, WTP= amount the respondent is willing to pay for restoring the Isahaya Bay Wetland in yen (under open-ended elicitation method). Support= a dummy variable representing whether the respondent support the IBRP (yes= 1 if they support 0 otherwise). Influence= the degree of influence the respondent felt to have received from IBRP and have five levels which rises with rise in degree of influence received. Visit = dummy variable representing whether the respondent went to visit the IBW at least once. Years residing = the number of years the respondent living in the present address. Income = annual income of the respondent's household divided by 10 million yen and considered as covariates as it has 14 categories. Age = age of the respondent. Sex = dummy variable representing respondent's gender.

11.5 Model Result and Interpretation

Best-subset analysis was conducted separately for the respondents of Isahaya, Nagasaki and Kitakyushu and the results are summarized in Table 11.2 and 11.3. All the variables mentioned above have *VIF*< 2, thus no fear of collinearity exists.

As shown in Table 11.2, finally three models are selected based on C_p statistic. For lack of space only the best models are shown here. All of these have C_p statistic<(k+1). Thus for Isahaya- *support*, *income* and *years residing*; for Nagasaki-*support*, *influence*, *income*; and for Kitakyushu- *influence*, *income* and *age* should be included in the model. Based on this, we have performed the regression analysis and found that the variables presented in Table 11.3 as statistically significant.

From the results of the regression analysis it can be observed that, in all the cities, the *influence*, *income* and *age* are positively related with WTP. On the other hand, *support* and *years residing* are negatively related to WTP. Among these, let us focus on the results of *influence* and *support* as these qualitative data are extracted from the free comments. Accordingly, for one level of increase in influence received form IBRP, the WTP increases by approximately ¥879 in Nagasaki and ¥805 in Kitakyushu. Whereas another qualitative factor *support* (whether the respondent support IBRP or not) coefficient indicates that, if the respondent is a supporter of IBRP then WTP decreases by approximately ¥1,665 in Isahaya and ¥2,856 in Nagasaki. This indicates that in Isahaya even if the respondent is a supporter of IBRP, the WTP does not fall as much as in Nagasaki. One explanation to this might be that, although a group of respondents in Isahaya are supporting IBRP, but they are also willing to pay for protecting the Isahaya Bay Wetland.

Table 11.2. Selected models by best- subsets analysis for reducing the number of explanatory variables

	Isahaya City	Nagasaki City	Kitakyushu City
Variables	Sup,Inc,Yrs	Sup,Inf,Inc	Inf,Inc,Age
Cp	2.23	3.53	2.96
k	4	4	4
R Square	0.1147	0.3414	0.1909
Adj. R Square	0.0744	0.3048	0.1535
Std. Error	5316.33	4614.01	5239.07

Table 11.3. Results of the regression analysis by using best- subsets approach to model building (Dependent Variable = WTP in yen)

Variables	Isahaya City Coefficient (t-stat) [n=58]	Nagasaki City Coefficient (t-stat) [n= 55]	Kitakyushu City Coefficient (t-stat) [n= 68]
Constant	3962.59 (2.253)	93.41 (0.058)	-9318.93 (-2.483)
Support (Sup)	-1664.95 (-1.198)[a]	-2856.24 (-1.746)[a]	–
Influence (Inf)	–	878.50 (1.982)[a]	805.37 (1.839)[a]
Income (Inc)	5.28 (2.157)[b]	6.58 (3.301)[b]	7.05 (3.383)[b]
Yrs.Res. (Yrs)	-27.68 (-1.20)	–	–
Age (Age)	–	–	146.59 (2.538)[b]

Note: [a] Indicates that the coefficient is statistically significant at, at least, the 0.10 level.
[b] Indicates that the coefficient is statistically significant at 0.05 level.

11.6 Summary and Conclusion

The findings of this chapter can be summarized and concluded through the following points:

1. From the graphical representation of the free comments of the respondents of different cities, it was evident that, there is a huge gap between the respondents of Isahaya compared to that of Nagasaki and Kitakyushu (see Figs. 11.4~11.9). On an average, the residents of Isahaya are supporting the IBRP for fear of flood disaster and water logging. On the other hand, the residents of Nagasaki and Kitakyushu are having a mixed feeling regarding the project with tendency to oppose the project for saving the eco-system.
2. From the results of the regression analysis we can perceive that, free comments contain different valuable information, which might not possible to extract by asking direct questions. Also the comments summarized can show pivotal relationship with WTP, which might not be possible to comprehend by viewing only monetary WTP value.
3. Finally, it is the responsibility of the researchers to summarize the comments of the respondents in an understandable way and to convey their independent thoughts to the society and concerned policy makers through published documents. This is also the expectation of the respondents spending his or her invaluable time to complete the survey questionnaire and this feeling was echoed in many free comments in case of the Isahaya Bay Wetland survey.

CHAPTER 12 Integration of CVM and Remote Sensing Technique for Environmental Valuation

12.1 Introduction

The environmental valuation made by using CVM has been questioned by many researchers. The way of asking question, response rate, unfamiliarity with the problem, behavior of the respondents etc. can make the valuation made by CVM biased. One suggestion which is put forward to as countermeasure to this problem is to verify the results of CVM by conducing valuation by other techniques. This chapter endeavors to do so by comparing the valuation made by CVM with valuation made by applying remote sensing technique. Thus the objectives of this part of the book are to conduct environmental valuation through the integration of remote sensing and CVM; and to see the difference in valuation of environmental impact by behavioral economic tools based survey and thematic mapping of land cover derived from satellite remote sensing images. Both the Isahaya Bay Wetland (IBW) and Huis Ten Bosch (HTB) case study database are used to conduct this verification.

12.2 Data and Methods

The research methodology followed in this chapter is explained in Fig. 12.1 and 12.2. For CVM valuation, database of IBW and HTB in Chapters 3 and 8, respectively are used. Whereas, for valuation based on remote sensing, satellite images of LANDSAT-5 before and after the initiation IBW and HTB projects are used for comparison. LANDSAT-5 launched on March, 1984 having ground resolution of 30 meter for all bands, except for band 6 with 120 meter resolution. Thus one pixel is equivalent to 900 m^2 for bands 1- 5 and 7. The selection of the satellite images were done strictly on the following criteria:

1. High quality images with minimal or no cloud coverage.
2. Identical time period.
3. During low tide period (below 2 meter).

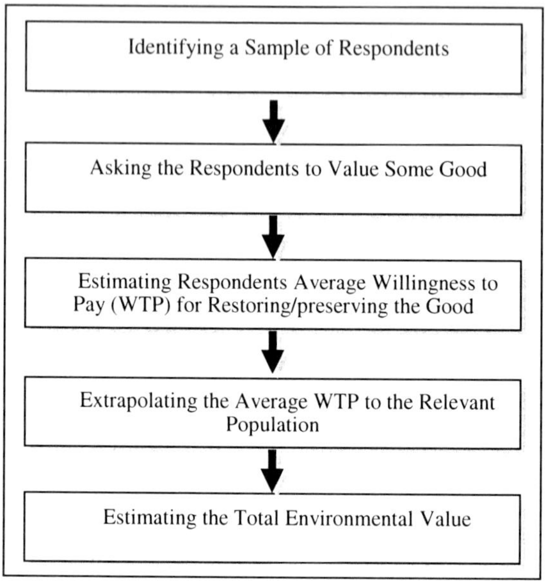

Fig.12. 1. Steps involved in environmental impact assessment by applying CVM

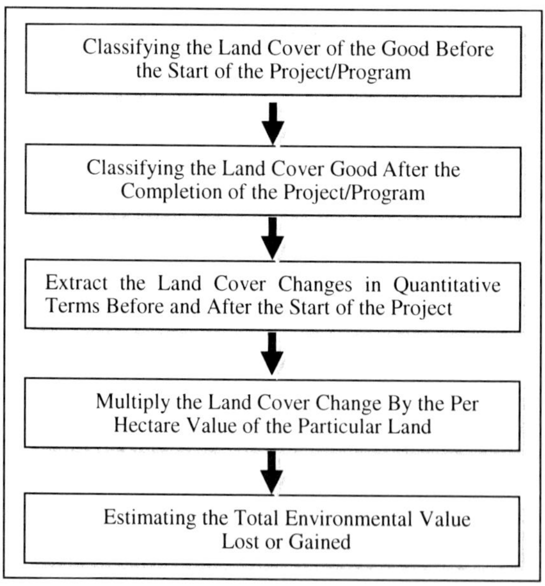

Fig. 12.2. Steps involved in environmental impact assessment by remote sensing

12.3 The IBW Case Study

12.3.1 Environmental Impact Valuation by CVM

The WTP estimated by applying Turnbull method in all the three cities revealed that the mean (WTP) is approximately ¥6,500 per household for all the sample respondents of Isahaya, Nagasaki and Kitakyushu (See Table 12.1). We are going to use this amount as a representative sum, and extrapolate it to the households of Kyushu. In doing so, we are going to divide the household population of Kyushu Island into two categories on the basis of use and non-use values for IBW. In the first category of use value, the entire households of four prefectures of Kyushu located on the coast of Ariake Sea- Nagasaki, Saga, Kumamoto and Fukuoka (NSKF) are taken which are repotted to be having direct impact due to the commencement of reclamation project on the IBW (see row 6 in Table 12.1). On the other hand, in the second category of non-use value, the households of remaining prefectures of Kyushu, viz., Oita, Miyazaki and Kagoshima are also included as although they are not having any direct impact from IBW, but possess potential optional conservation value for it (see row 7 in Table 12.1).

Accordingly, after extrapolating the mean WTP of ¥6,500 to the respective category of households, it was found that, the residents of directly affected NSKF prefectures and residents of entire Kyushu are willing to pay approximately ¥22 billion and ¥32 billion, respectively, to restore the Isahaya Bay Wetland (See Table 12.2). This amount would be considered as the monetary value of the negative impact caused to IBW, as the respondent's are willing to pay this amount to restore the ecological condition of the region to the position it was before the initiation of IBRP.

12.3.2 Environmental Impact Valuation by Land Cover Mapping

Estimating the Area of Wetland Destroyed

In order to estimate the portion of the wetland destroyed by the IBRP, the land cover changes are examined by satellite images of LANDSAT/TM 5 before and after the initiation of the project. Although the sea dikes of the project are closed in April 1997, the construction work in the wetland area began long before. Hence, the images of April 1988 were selected to see the extent of wetland before any sort of project work begun (see Table 12.2). This was then compared with that of April 1999 to see the extent of wetland destroyed since the initiation of the project work. Figure 12.3 shows such land cover changes result by unsupervised classification. Only the water area is extracted to have more insight into the wetland

Table 12.1. Total estimated WTP

1. Sample size (complete)	396
2. Mean WTP (¥/household)	6,500[a]
3. Median WTP (¥/household)	4,000
4. SD of the Mean	399
5. Range of 95% confidence interval (¥)	± 1,565.3
6. Total No. of Households in NSKF	3,372,400[c]
7. Total No. of Households in Kyushu	4,976,000[c]
8. Estimated Total WTP for NSKF (2X6)	21.92[b]
9. Estimated Total WTP for Kyushu (2X7)	32.34[b]

Note : [a] Rounded to nearest 100 yen.
[b] Billion yen.
Source: [c] MPHP 2003.

Table 12.2. Images used

Date	Status of the project
April 15, 1988	Before the initiation of IBRP
April 14,1999	After closing off the sea dikes

area changes. From the results of the classification verified with ground truth, it has been reveled that in April 1988 there had been approximately 4,000 ha of wetland in the region. Conversely, in April 1999 it is not at all exaggerating to say that, the 3,550 ha (the area in the left side of the sea dikes as shown in the image a of Fig. 12.3) of wetland out of 4,000 ha cut off by the reclamation project have been totally destroyed. The cut out area, which was previously wetland, are now turned into shallow water or catch basin (62%), farm land or soil (34%) and sea dike or concrete structures (4%) (see image b) of Fig. 12.3). Thus, both by land cover images and ground truth it has been confirmed that, around 3,500 ha of wetland area enclosed by IBRP by shutting down sea dikes has been destroyed.

Monetary Impact Estimation by Land Cover Mapping

In this section, we have attempted to convert the 3,500 ha of the Isahaya Bay Wetland destroyed into monetary figures. In doing so, we need to have the per hectare value of the wetland. The usual procedure to estimate the per hectare value of the wetland is to add the monetary benefits of the most important functions provided by wetland, such as: flood prevention, storage and recycling of human waste, nursery value, aquaculture, recreation, food production, education and science use, etc. For the Dutch Wadden Sea adding these values amounts to a value of \$6,200 ha^{-1} yr^{-1} (de Groot 1992). Surely, values for other wetlands in different countries would vary and compared to other valuation of wetlands this amount is rather a moderate estimate. For example, Gosselink *et al.* (1974) calculated a value of \$10,000 ha^{-1} yr^{-1} for the wetland in the east coast of the United States. Among

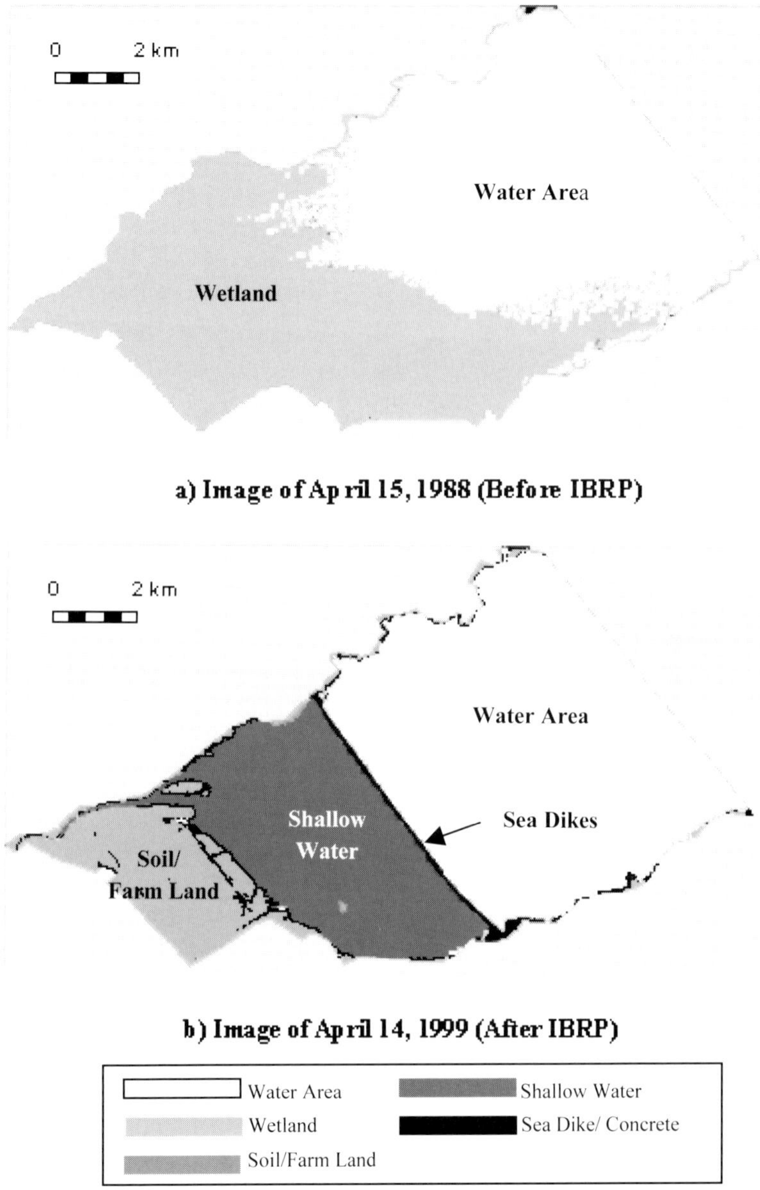

Fig. 12.3. Land cover changes in the IBRP site form the images of LANDSAT-5

Table 12.3. Wetland services and their global value

Ecosystem services	$\$ha^{-1}$ $year^{-1}$
Gas regulation	133
Disturbance regulation	4,539
Water regulation	15
Water supply	3,800
Waste treatment	4,177
Habitat/Refugia	304
Food Production	256
Raw Materials	106
Recreation	574
Cultural	881
Total Value	14,785

Source: Costanza et al. (1997)

higher estimates, Thibodeau and Ostro (1981) arrived at a value of \$28,000 ha^{-1} yr^{-1} for the Charles River Basin in Massachusetts. In another estimate placed the economic value of a hectare of Atlantic Spartina Marsh at over \$72,000 ha^{-1} yr^{-1} (Hair 1988). On the other hand, in a more detailed study, Costanza et al. (1997) attempted to place a total value on the Earth's ecosystem and estimated the total area covered by 17 biomes classified by Bailey. In valuing each biome, the services provided are identified and given a monetary value based on past research and original calculations. While valuing wetland, Costanza et al. (1997), identified ten ecosystem services of wetland and accordingly placed a value for each (see Table 12.3). The value of wetland was estimated as 1.72 million yen ha^{-1} yr^{-1} (\$13,784 ha^{-1}, converted by taking \$1=¥115.20 as on 17 September 2003).

Therefore, there are various studies concerning the valuation of wetland providing value ranging from \$6,200 ha^{-1} yr^{-1} to \$72,000 ha^{-1} yr^{-1}. By considering the fact that there is no comprehensive study estimating per ha value of wetland in Japan, in this study we have taken the estimation made by Costanza et al. (1997), of 1.72 million yen ha^{-1} yr^{-1} value of wetland for multiplying with the 3,500 ha of the Isahaya Bay Wetland lost for the following merits:

1) Summarized the wetland value of various studies;
2) Considers a wide range of services provided by wetlands;
3) Applied both willingness-to-pay (WTP) and energy analysis (EA) valuation;
4) Provides a more generalized, moderate and conservative value for wetlands throughout the world.

Table 12.4. Present value of wetland lost

	Discount rate	
	3%	8%
Wetland Area Destroyed (ha)	3,500	3,500
Wetland value ha^{-1} $year^{-1}$ (million ¥)	1.72	1.72
Present value of wetland ha^{-1} ((million ¥)	57.33	21.50
Total estimated present value of wetland lost (billion yen)	200.66	75.25

Table 12.5. Summary of wetland impact value estimates

	Impact Valuation Method	
	CVM (Billion yen)	Land Cover Mapping (Billion yen)
Lowest Value	22	75
Largest Value	32	200

Based on this, we have calculated the present value of wetland per ha in perpetuity by taking 3% and 8% as the discount rate range to see the variation, though a lower discount rate is preferred for wetland valuation (Costanza et al, 1989). The formula for calculating present value of a sum in perpetuity is as follows:

$$PV = \frac{C}{r} \tag{12.1}$$

Where, PV is the present value, C is the annual cash value and r is the discount rate. The present values calculated are shown in Table 12.4, provides us present value of wetland lost ranging form ¥75 billion to ¥200 billion depending on the discount rate.

12.3.3 Discussions

The ongoing environmental impact assessment by applying two different valuation approaches can be summarized in the following way:

1. Under CVM, the extent of economic value lost due to the initiation of IBRP on IBW is within the range of ¥22 to ¥32 billion (see Table 12.1). The value would even rise further if we extrapolate the average WTP to the entire household population of Japan. Hence this estimate is rather a conservative one.
2. While under land cover changes of remote sensing, the extent of economic value lost due to the initiation of IBRP on IBW is within the range of ¥75 billion to ¥200 billion (see Table 12.4). This estimate is also from

conservative point of view as we have used a moderate per hectare value of wetland.

3. If we compile the two different impact valuation methods, then we can conclude that the environmental value lost due to the destruction of the Isahaya Bay Wetland by the reclamation project is worth ¥22 billion at the lowest value and ¥200 billion at the largest value (see Table 12.5). Some where in between these or beyond them the true value lies.

12.4 The HTB Case Study

12.4.1 Environmental Valuation by CVM

By applying Contingent Valuation survey study, the mean (WTP) is estimated to be approximately ¥4,000 per household for all the sample respondents of Sasebo and Nagasaki Cities. (see Table 12.6). The median (WTP) is ¥2,000 per household. Next we are going to extrapolate the mean WTP of ¥4,000 to the entire 4,976,000 (MPHP, 2003) private households of Kyushu to estimate the benefits received from various environmental costs incurred by HTB. Accordingly, we found that, the residents of Nagasaki Prefecture are willing to pay approximately ¥2.17 billion to preserve the Huis Ten Bosch.

12.4.2 Environmental Valuation by Land Cover Mapping

Examining the Land Cover Before and After the HTB Project

In order to see the environmental impact of the HTB project, we have examined the land cover changes by satellite images of LANDSAT-5 before and after the initiation of the project. Although HTB was officially opened for public in 1992, its construction work began in 1988. Hence in this study we have taken the images of April 1988 to see the land cover position of the HTB project area before any development work started. Then we have compared it with that of April 2000 to see the change in the land cover due to various activities of HTB. Table 12.7 and Fig.12.4 show such land cover changes result by unsupervised classification. From the results of the classification verified with ground truth, it has been revealed that in 1988, the HTB project site was mostly industrial wasteland consisting of about 191 ha of barren land and unhealthy grass, and about 80 ha of demolished concrete structures. Thus the land cover of the area was mostly unproductive industrial wasteland (see Fig. 12.4a). On the other hand, land cover classification

Table 12.6. Total estimated WTP

Description	Number/ Amount
Sample size (complete)	479
Mean WTP (¥/household)	4,000 [a]
Median WTP (¥/household)	2,000
SD of the Mean	238
Range of 95% confidence interval (¥)	± 467
Total Number of Households in Nagasaki Prefecture	542,985 [b]
Estimated Total WTP (Billion yen)	2.17

Note : [a] Rounded to nearest 100 yen.
Source: [b] MPHP, 2003.

in 2000 shows that land cover in the area changed a lot with transformation of lands into water/ lake, forest, grass and rangeland and building structures (see Fig. 12.4b). Thus, both by land cover images and ground truth it has been confirmed that, HTB project contributed positively towards restoring the eco environment in the area.

Monetary Valuation by Land Cover Mapping

In this section, we have attempted to convert the positive environmental impact (benefits) resulted from HTB project into monetary figures. In doing so, we need to have the per hectare value of these impacts. Costanza et al. (1997) attempted to place a total value on the Earth's ecosystem and estimated the total area covered by 17 biomes classified by Bailey. The value placed for water/ lake, forest, grass and rangeland in this study are ¥934 thousand ha^{-1} yr^{-1}, ¥106 thousand ha^{-1} yr^{-1} and ¥26 thousand ha^{-1} yr^{-1}, respectively (converted by taking $1=¥110) (see Table 12.8).

Based on this, we have calculated the present value of all these three environmental goods in perpetuity by taking 3% as the discount rate, as lower rate is usually preferred for environment friendly public beneficiary projects. The formula for calculating present value of a sum in perpetuity is same as Eq. 12.1. The present values calculated are shown in Table 12.9 and provide us present value of total environmental benefits provided by HTB as ¥1.58 billion at 3 percent discount rate.

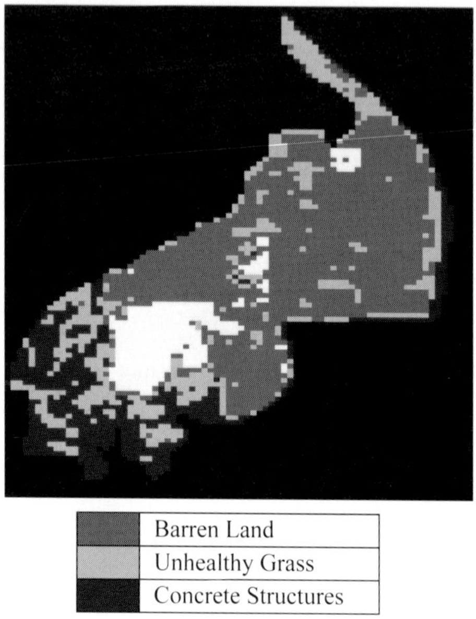

(a) 1988 (Before HTB project)

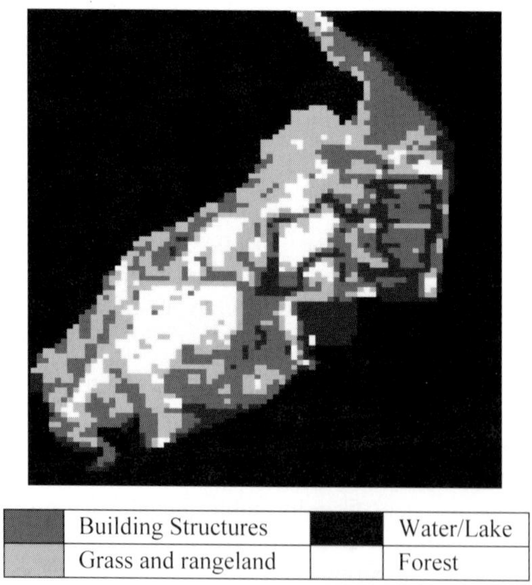

(b) 2000 (After HTB project)

Fig. 12.4. Land cover changes in the HTB area

Table 12.7. Land cover change in the HTB area

Classification	1988 Area in ha	2000 Area in ha
Water/Lake	-	43
Forest	-	46
Unhealthy Grass	53	-
Grass and rangeland	-	85
Barren Land	138	-
Concrete Structures	80	-
Building Structures	-	93

Table 12.8. Environmental benefit analysis of HTB

	Area (ha)	Value $ha^{-1} yr^{-1}$ (¥)	Total Value yr^{-1} (Mil. ¥)
Water/Lake	43	934,000	40.16
Forest	46	106,000	4.88
Grass and rangeland	85	26,000	2.21
Total			47.25

Table 12.9. Present value of environmental benefits

	Discount Rate 3%
Total Value of Env. Benefits $year^{-1}$ (million ¥)	47.25
Present value of Benefits (million ¥)	1575
Total estimated present value of Env. Benefits (billion ¥)	1.58

Table 12.10. Summary of environmental valuation estimates

Valuation Method	Value (Billion ¥)
CVM	2.17
Land Cover Mapping	1.58

12.4.3 Discussions

By compiling both of the valuation results we can conclude that, the value of preserving HTB is in between 2.17 billion yen to 1.58 billion yen (see Table 12.10). Somewhere near this the true value lies. Valuation of HTB is very much timely as presently rehabilitation activities are going on in HTB after financial bankruptcy. This study in this regard is merely a modest attempt to integrate CVM and satellite remote sensing techniques and leaves the ample scope for further studies and discussion.

12.5 Conclusion

Estimation of monetary value of wetland lost due to some human activities is a very difficult and complex task. But it is essential for safe guarding the invaluable environmental resources from being vanished and thus demands proper environmental impact assessment. This chapter in his regard is a modest attempt to integrate CVM and satellite remote sensing techniques not only to judge the impact qualitatively, but also to monetarily quantify the negative impact caused. The value estimated from conservative viewpoint and expected to provide some concrete base to community and stakeholders to make their decision regarding environmental impact caused to the Isahaya Bay Wetland by IBRP and on issues relating to the sustainability of HTB. For deriving values the authors had to rely on generalized global wetland valuation estimates in case of remote sensing. But this is largely due to the absence of definite valuation studies particular to the IBW and HTB. Accordingly this area leaves the scope for detailed further studies and discussion.

CHAPTER 13 Effect of Question Format on WTP in CV Studies

13.1 Introduction

As it is well known, there are different ways to ask willingness to pay (WTP) questions in contingent valuation (CV) surveys, which are known as elicitation methods. Presently four types of elicitation methods are commonly used in CVM studies e.g., open-ended (OE), bidding game (BG), single-bound dichotomous-choice (1DC) and double-bound dichotomous-choice (2DC). Among them especially dichotomous-choice (DC) format is the most widely used one. Follow-up questions are also used to increase the precision of the estimate with DC question. The NOAA blue ribbon panel advocated this method as the most appropriate one in most circumstances (Arrow et al. 1993). While among others bidding game (BG) approach has been almost deserted because it tends to result in a starting point bias. However, all these methods of asking questions have their relative advantages and disadvantages and none is free from criticisms. In this context, this study is a modest attempt to study the impact of different elicitation methods on willingness to pay of the same respondents by using the CVM study data of two environmental assessment case studies.

Table 13.1. Description of the survey

	HTB Study	Public Park Study
Location	Nagasaki City	Nagasaki City
Questionnaire Distributed	950	1000
Questionnaire Returned	232	194
Elicitation Methods Used	OE, 1DC, 2DC	OE, 1DC, 2DC

13.2 Data and Methods

The database of the HTB and public park case studies are used to examine the effect of elicitation methods on the WTP of the respondents (see Table 13.1 for survey details). Both of these studies have Nagasaki City as their sample area and attempted to estimate the value of sustaining and increasing recreational parks.

As a methodology statistical model has been constructed by dividing the elicitation methods into two groups OE and DC responses. 1DC and 2DC are merged into a single model of DC, as the same 1DC model can be extended to 2DC with the basics remaining the same (Bateman et al. 1999). Also the 'anchoring effect' (the probability of the respondent accepting the offered bid) is common in all DC formats. Thus 2DC, 3DC etc. are a simple extension to the DC bidding game and therefore the drawbacks of IDC also applies (Hanemann et al. 1991; McFadden and Leonard 1993).

13.3 Findings

The findings of the analusis are summarized through Table 13.2 and Figs. 13.1 and 13.2.

13.3.1 Open-ended Format (HTB Study)

Under the OE elicitation method the mean WTP is estimated as ¥2,760 for the entire sample (95% CI=¥2,150~¥3,370) (see Table 13.2). All those who refused to pay anything are treated as having zero WTP. But as common to the OE format the standard deviation is revealed to be very high. A model is constructed to check the validity and explanatory power of the related variables. Accordingly, the double-log form model derived is as follows:

$$\underset{}{\text{LWTP (OE)}} = \underset{(7.192)}{5.854} + \underset{(1.547)}{0.219}\text{LINC} \quad (13.1)$$

Where,

LWTP (OE) = Natural log of open-ended WTP response
LINC = Natural log of the respondent's income
Numerals in the parenthesis are *t* values.

The variables shown above are significant at least at 0.10 level. Other variables such as: visit and demographic variables sex and age were also inserted in the model, but they were not significant. The R^2 (adj.) = 4%. Thus it indicates that the explanatory variables are not properly explaining the OE willingness to pay.

13.3.2 Dichotomous-choice Format (HTB Study)

Under the DC elicitation method the mean WTP is estimated as ¥3,360 (95% CI= ¥2,700-¥4,020) (see Table 13.2). The standard deviance reduced significantly compared to the OE format. With dichotomous structure of the dependent variable, the model can be estimated through a non-linear probability model and the most common one is the logit model (Pate and Loomis 1997). Accordingly a logistic model is constructed to check the validity and explanatory power of the relevant factors is as follows:

$$Ln[prob(yes)/\{1-prob(yes)\}]=6.670-0.6272LBID \qquad (13.2)$$
$$(10.686) \quad (10.688)$$

Where,

LBID = Natural log of the proposed bid value
Numerals in the parenthesis are t values.

The variables shown above are significant at least at 0.10 level. The overall explanatory power of the model is higher than OE format. The overall model is significant at the 0.01 level. The model reasonably predicts 71% of the responses correctly. The R^2 (adj.) = 8.88%. Thus it indicates that the explanatory power of the variables increased in the DC format. But the influence of the bid level variable remains highest, indicating the presence of anchoring effect, as it alone can explain almost 71% changes in the willingness to pay.

13.3.3 Open-ended Format (Public Park Study)

In this study, OE format mean WTP is estimated as ¥4,660 for the entire sample (95% CI=¥3,880~¥5,440) (see Table 13.2). Also like the HTB study, the standard deviation is high. A model is constructed to check the validity and explanatory power of the related variables. Here the double log form model derived is as follows:

Table 13.2. Mean WTP under various elicitation methods

Elicitation Method	HTB Study			Public Park Study		
	Mean WTP (¥)	SD	95% CI (¥)	Mean WTP (¥)	SD	95% CI (¥)
OE	2,760	4,230	2,150-3,370	4,660	5,490	3,880-5,440
1DC	3,360	330	2,700-4,020	4,960	240	4,490-5,430
2DC	3,520	320	2,890-4,150	5,370	340	4,700-6,040

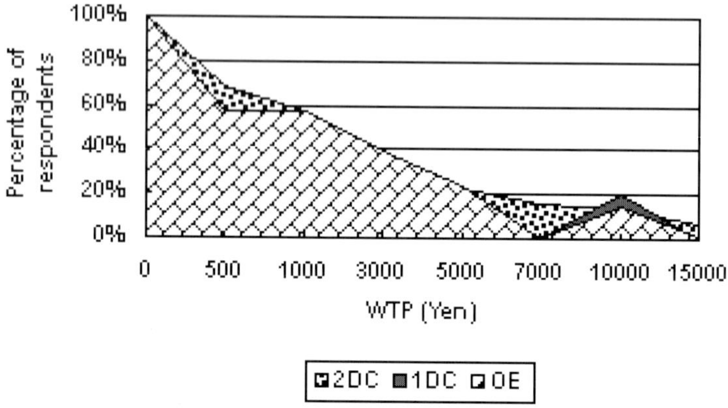

Fig. 13.1. Effect of elicitation method on WTP for HTB case study

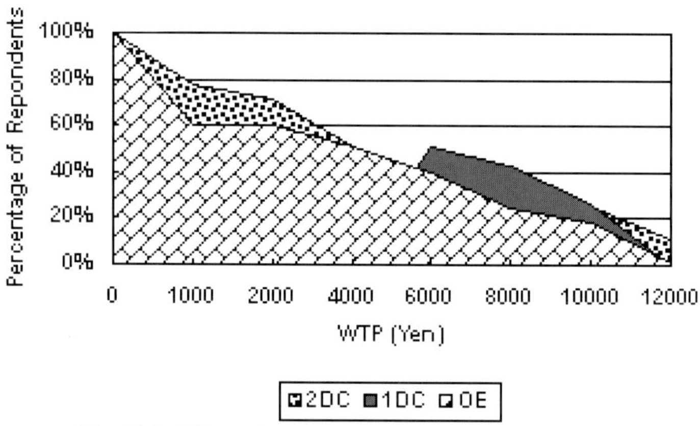

Fig. 13.2. Effect of elicitation method on WTP for public park case study

$$\underset{}{\text{LWTP (OE)}} = \underset{(7.321)}{7.781} + \underset{(1.761)}{0.285\text{VISIT}} + \underset{(0.571)}{0.003\text{AGE}} + \underset{(0.378)}{0.059\text{LINC}} \tag{13.3}$$

Where,

LWTP (OE) = Natural log of open-ended WTP response
VISIT = Whether the respondent went to visit public parks at least once in a month. Yes=1, No=0
AGE = Age of the respondent's in years
LINC = Natural log of the respondent's income
Numerals in the parenthesis are *t* values.

Table 13.3. Aggregate willingness to pay (In million yen)

Elicitation Methods	HTB Study	Public Park Study
OE	486	820
1DC	591	873
2DC	619	945

Note: Average WTP of Table 13.2 is multiplied by the total number of households of Nagasaki City (MPHP 2003).

The variables shown above are significant at least at 0.10 level. But the overall explanatory power of the model is as usual very low. The R^2 (adj.)= 0.5%.

13.3.4 Dichotomous-choice Format (Public Park Study)

Under the DC elicitation method the mean WTP is estimated as ¥4,960 (95% CI= ¥4,490-¥5,430) (see Table 13.2). The standard deviance reduced highly as compared to the OE format. The logistic model constructed to check the validity and explanatory power of the relevant factors is the following:

Ln[prob(yes)/{1-prob(yes)}]=

$$1.883 - 0.665LBID - 1.290VISIT + 0.054AGE + 0.002INC \quad (13.4)$$

(0.485) (5.226) (14.011) (11.823) (12.335)

Where,

LBID = Natural log of the proposed bid value
INC = Income of the respondent household (in ten thousand yen)
Numerals in the parenthesis are *t* values.

Rest of the variables are the same as explained before. The variables shown above are significant at least at 0.10 level. The overall explanatory power of the model is higher than OE format. The model reasonably predicts 71.94% of the responses correctly. The R^2 = 15.99%. As also seen in the HTB study, influence of the bid level variable remains highest, indicating the presence of anchoring effect, as it alone can explain almost 60% changes in the willingness to pay.

13.4 Discussion and Conclusion

From the findings of this chapter the following points can be put forward:

1. From the results of both of the case studies, we have seen that under OE question format willingness to pay (WTP) is lower and the variance is higher leading towards the possibility of underestimation of the true WTP (see Table 13.3 and Figs. 13.1-13.2).

2. Whereas under dichotomous-choice question format the WTP is higher and the variance also decreases.
3. As shown in Table 13.3, extrapolating the average WTP calculated to the total number of households of Nagasaki City makes the impact of elicitation methods on aggregate environmental assessment more clear. The range of variation is in between ¥486 million to ¥619 million in HTB study and ¥820 million to ¥945 million in public park study. This indicates the larger impact of choosing elicitation method on environmental assessment. But the difference is not as significant as evident in other relevant studies (see Bateman et al. 1999).
4. From the model construction statistical inaccuracy of the OE format is again revealed due to poor ability of the explanatory variables to explain the WTP.
5. DC format model construction showed more relative explanatory power of the related variables, but highly influenced by the bid level.

From the findings of this chapter we have seen that, both OE and DC approach have their relative advantages and disadvantages. Lack of knowledge of cost-benefit and free riding lead to understatement of WTP under OE question format. Whereas, although DC format is free from many of the demerits of OE format, still the 'framing' or 'anchoring' effect arising from the probability of accepting the bid level due to ignorance about true valuation is present. Thus to sum up, both OE and DC format are prone to biases and may lead to improper environmental assessment, as the respondents might base their answers without having adequate knowledge about the actual cost-benefit behind the valuation. In order to avoid this, we would like to propose an alternative method, which are expected to overcome the demerits of the two above stated methods.

As shown in Fig. 13.3, in the proposed method additional information on cost-benefit of the concerned good and other relevant good would be provided in order to assist the respondent's decision in numerical from. Also a number of randomly assigned bid values would be provided to reduce anchoring effect arises under DC format. Accordingly, it would be of great interest to test the accuracy of this proposed alternative question format in future environmental assessment studies.

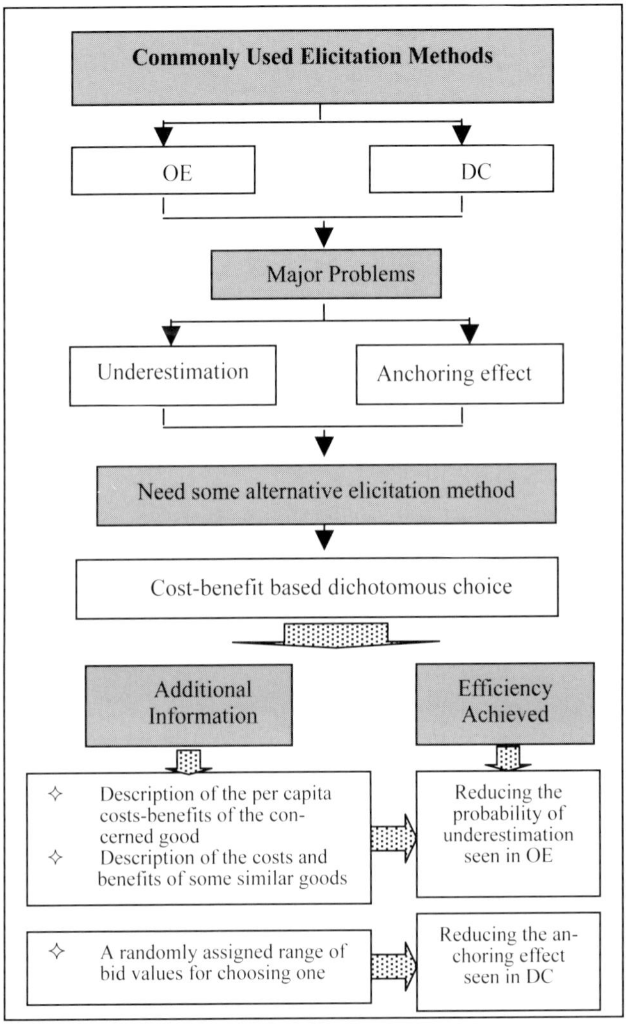

Fig. 13.3. Alternative elicitation method

PART VI: SUMMARY AND CONCLUSION

CHAPTER 14 Policy Recommendations and Conclusion

14.1 On the Case Studies

The case studies discussed in this book have several implications on the society and policymakers and can be translated into policy recommendations as follows:

1. The total willingness to pay (WTP) calculated might be treated as a proxy value of the amount of damages caused to the residents by the Isahaya Bay Reclamation Project (IBRP).
2. The WTP values calculated by using the contingent valuation method can provide a base for paying compensation to the potential affected interested groups by the concerned authorities.
3. Recognition of socio-environmental cost into cost-benefit analysis is very important to allow citizens and the concerned public body to decide on the viability of the project. Contingent valuation method (CVM) is recommended as one of the alternatives to insert socio-environmental considerations into the cost-benefit analysis.
4. Sensitivity analysis should be included in the cost-benefit evaluation process to judge the validity of the project in different degrees of uncertainty. As shown in the case of IBRP, various predictions were made about the future, which later on did not remain constant, such as, price of the construction materials, project plan, etc.
5. Regarding the IBRP it can be concluded that, undertaking development works to protect from natural disaster is definitely important, but careful long-term viability study should be conducted to ensure that it does not eventually destroy the invaluable and non-restorable eco-system.
6. The estimated aggregate WTP of ¥920 million derived in public park case study might be considered as an amount which if the policy makers spend to develop public parks in Nagasaki City, would increase the utility of the residents of Nagasaki instantly, as they are already willing to pay for it.
7. Necessary steps should be taken to increase presence of natural environmental in public parks and reduce artificial setups, which would in turn increase the number of visitors in public parks.

8. As balanced *living environment maintenance* by developing more public parks is becoming one of the top agendas for both the local and central government in Japan, the WTP amount derived from this study might be used as a base for financing the development of public parks.
9. More information dissemination is necessary to introduce the environment friendly initiatives taken by HTB. As our survey study revealed that even the residents of Sasebo City are not familiar about them.
10. A large number of the respondents felt that the entrance fee, attraction fee, hotel rent and foods are expensive inside HTB. Thus, necessary action should be undertaken to make HTB more cost effective. The following measures are suggested:
    a) In order to increase the number of repeaters, the entrance fee should be kept as minimum as possible. Thus, small sales margin with large turnover is suggested.
    b) There might be special discount group prices for the pupils or students of primary, secondary school and university and also for the general public to increase awareness on environmental issues.
11. More publicity and coordinated efforts are needed to convince people to recognize HTB not only as a theme park, but also as an environmental amenity which needs to be protected. HTB should make its purpose clear to the public, whether it is a theme park or an environmental amenity. Because, if it is a theme park, it should survive by its own efforts and must face the competition usually prevalent in a market economy. However, if it is an environmental amenity, then it can gain support of the general people for ensuring its existence and thus have different valuation considerations.

14.2 On Environmental Valuation Methodologies

For using contingent valuation method (CVM) for conducting cost-benefit analysis of environmental goods the following points might be suggested in order to get more steadfast results (also shown by the pyramid in Fig. 14.1):

1. Extensive pretesting study should be conducted before finalizing the CV survey instruments.
2. Various elicitation methods (format of asking CVM questions) should be used to compare the change in willingness-to-pay and their significance. This will improve the reliability of the study.
3. Developing models through multivariate analysis is very rewarding, as it would enable to identify the most important variable influencing willingness-to-pay.
4. If possible, CVM studies should be supported by valuation made by applying other techniques (e.g., travel cost method, remote sensing). This would guide in reaching towards the true range of environmental valuation for facilitating cost-benefit analysis and provide an opportunity to test the reliability of CVM study.
5. A space should be provided at the back of the CVM survey instruments, so that the respondents can write their comments freely regarding the problem under

consideration. This is highly recommended for conducting CVM surveys, especially in Japan, as results of our analysis of the case studies showed that these comments comprise valuable information, which is difficult to extract by only considering the monetary WTP related questions.

Finally, the case studies discussed in this book are merely some representative environmental cases in Japan. A very modest attempt has been taken though out this book to discuss the cases by applying contingent valuation method. The result of the studies and subsequent discussion are expected to provide the reader an insight into the approach and mindset regarding environmental cost-benefit analysis in Japan. However, replicating and synthesizing of the results revealed by the studies discussed in this book needs more comprehensive studies. In this respect this book is open for discussion

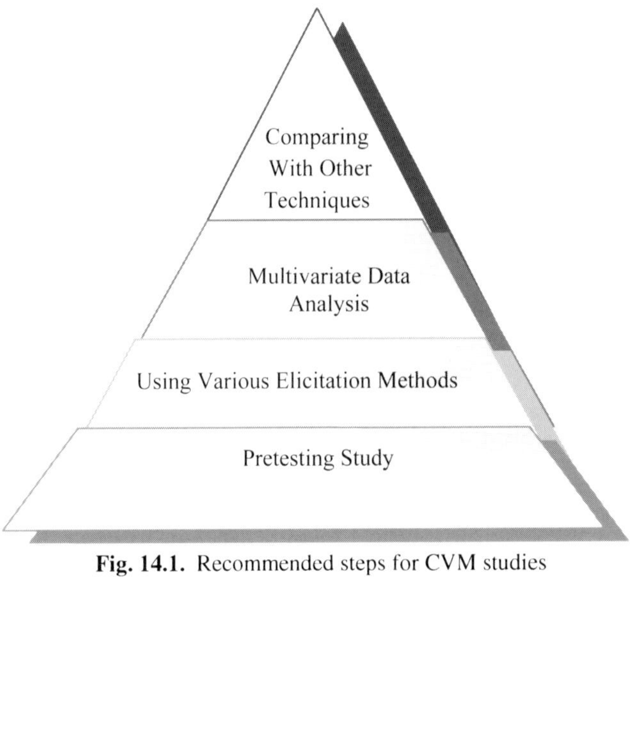

Fig. 14.1. Recommended steps for CVM studies

Appendix A The Isahaya Bay Wetland Study Questionnaire

The Nagasaki University and the Kyushu Kyoritsu University are jointly conducting a research survey concerning the 'environmental valuation of the Isahaya Bay wetland'. All the recipients of this questionnaire are selected randomly from the published telephone directory and the replies are anonymous. The data would be only used for the purpose of research. We would be very grateful if you kindly spare a few minutes of your time to answer this questionnaire.

A. Scenario Description

The Isahaya Bay Wetland, located in the Ariake Sea of Nagasaki Prefecture, is one of the biggest tidal flat wetlands of Japan. Various marine species including mudskipper and about 144 types of water birds used to live here.

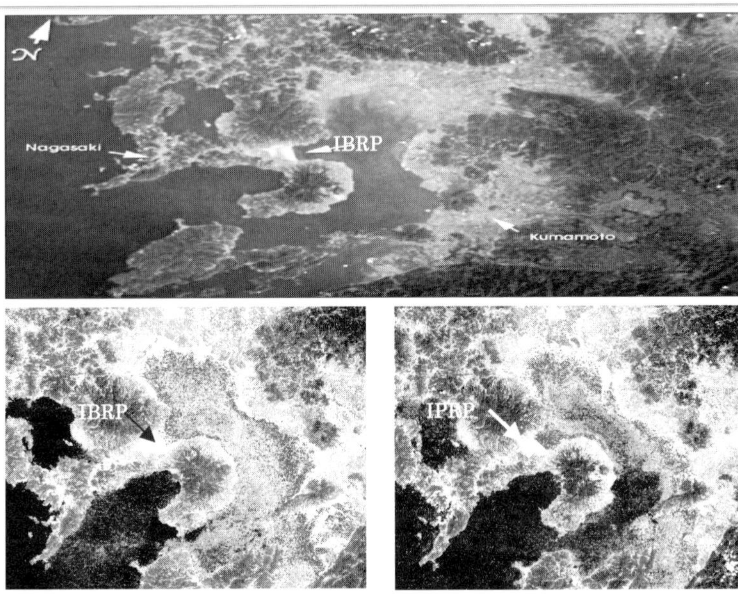

April 1999 **May 1986**

Satellite images showing the location and changes in the environment of the Ariake Sea

As we know, tidal flat provides various benefits e.g., water quality control, food production, nursery value, disaster prevention, habitat of aquatic species, recreation, storage and recycle of waste, stabilization of sediments etc. According to a survey of the Japanese Ministry of Environment in 1992, the estimated tidal flat wetland in Japan was about 514 km^2 among which the Ariake sea contains about 207 km^2 (40%).

However, recently this Isahaya Bay Wetland came into a lot of discussions due to the construction of sea dikes on it and closing off 3,550 ha of the bay by a 7 km long seawall. This project is blamed to have destroyed tidal flat area of about 15 km^2 which is about 3% of the tidal flat area of Japan and 7 % of the Ariake Sea. Local fishermen blame the reclamation project for serious ecological disaster in the region. Seaweed farmers allege that since a set of dikes sealed off part of the bay for reclamation in 1997, the sea has been polluted by an inflow of nutrient-rich waters, resulting in a poor seaweed crop.

Thus, through the reclamation project not only the ecology of the Isahaya Bay Wetland but also the total ecological balance of the Ariake Sea is threatened. We should all think about this problem seriously once again.

B. Questions Regarding Association with the Wetland

1 Did you go to visit the Isahaya Bay Wetland? If yes, then how many times? Please choose one from the following and encircle.

1. Visited within last one year(______times)
2. Visited one year before
3. Did not visited
4. Don't Know

2 How familiar are you with the following terms? Please encircle one from the each category.

The Isahaya Bay	1. Familiar	2. Heard only the name	3. Unfamiliar
Mudskipper	1. Familiar	2. Heard only the name	3. Unfamiliar
Wetland	1. Familiar	2. Heard only the name	3. Unfamiliar
Ecology	1. Familiar	2. Heard only the name	3. Unfamiliar

3 What is the image of the Isahaya Bay reclamation project (IBRP) to you? Please encircle from 1-6 of the each categories.

Image	Feel Strongly	Feel a little	Moderate	Don't feel that much	Don't feel at all	Don't Know
Needed for agricultural land reclamation	1	2	3	4	5	6
Needed for protection from flood	1	2	3	4	5	6
Needed for regional development	1	2	3	4	5	6
Affecting fishing industry	1	2	3	4	5	6
Affecting ecosystem	1	2	3	4	5	6
Affecting scenic-view	1	2	3	4	5	6

4 How the IBRP is affecting your daily life? Please encircle one of the following:

1. Influencing a lot
2. Influencing a little
3. Not influencing that much
4. Not influencing at all
5. Don't know

5 What sort of influences you are having from IBRP? Please encircle one of the following:

1. Income declined
2. Food quality deteriorated
3. Recreation spot reduced
4. Psychological influence
5. Environmental damage
6. No influence
7. Others (Please specify) ....

C. Environmental Valuation Questions

6

As mentioned above, a variety of living things inhabit the Isahaya Bay Wetland and formed an invaluable natural ecosystem. It is also the production place of marine resources, such as seaweed, shell etc. If the Isahaya Bay land reclamation project progress as it is, there is danger that these marine resources would be destroyed and such negative ecological impact would also expand to the entire Ariake Sea.

Now suppose we are going to restore the ecological environment of the Isahaya Bay Wetland as it was before the initiation of the reclamation project. This will cost us a considerable amount of money. For this we are going to create a fund named 'Isahaya Bay Wetland Preservation Fund' to restore the eco-system of the IBW.

Would you be willing to contribute ¥1,000 (assigned randomly ranging from ¥1,000 to ¥15,000) to this fund only once, to protect the IBW?

(Please answer by taking into account that this contribution will reduce your family budget)

Please encircle any one of the following:

1. **Agree** ⟹ **Please move to question 6-1**
2. **Disagree** ⟹ **Please move to question 6-2**

6 − 1

If you have said that you are willing to pay, is this because you are (please encircle the following):

1. Willing to protect the natural environment.
2. Willing to restore precious recreation spot.
3. Willing to protect wetland and eco-system.
4. Possessing desire to make future visit.
5. Willing to protect the natural environment for the future generation.
6. Others (Please specify) ....

Again if the amount is raised to ¥3,000 would you still be willing to pay for restoring the Isahaya Bay Wetland?

1. **Agree**
2. **Disagree**

6－2

Now if the amount is lowered to ¥700 would you be willing to pay for restoring the Isahaya Bay Wetland?

1. Agree

2. Disagree

If you have said that you are not willing to pay, is this because you are (please encircle the following):

1. Unwilling to pay as a fund.
2. Not interested.
3. Thinking that the Isahaya Bay is not valuable.
4. Feeling that the proposed amount is high. I can pay maximum ....
5. Feeling that the agricultural land and disaster prevention is more important. ....
6. Others (Please specify) ......

D. Question Regarding the Respondent

7

Please encircle your gender and number of family members:

1. Male 2. Female

Adult____persons Child (below 18 years of age)____persons

8

Please encircle your age class from the following:

① Below 20	② 21-30	③ 31-40	④ 41-50
⑤ 51-60	⑥ 61-70	⑦ 71-80	⑧ Above 80

9

Please encircle your job category from the following:

① Agriculture ②Forestry ③ Fishing ④ Manufacturing industry ⑤ Construction industry ⑥ Transportation ⑦ Wholesale/ retail trade ⑧ Finance and real estate business ⑨ Tourist business ⑩ Service industry ⑪ Government official ⑫ Teacher ⑬ Housewife ⑭ Student ⑮ Unemployed ⑯ Others

10 What is the gross the annual income of your household (Please encircle any one of the following):

1) Below ¥2million	2) ¥2-3 million	3) ¥3-4 million
4) ¥4-5 million	5) ¥5-6 million	6) ¥6-7 million
7) ¥7-8 million	8) ¥8-9 million	9) ¥9-10 million
10) ¥10-11 million	11) ¥11-12 million	12) ¥12-13 million
13) ¥13-14 million	14) ¥14-15 million	15) Above ¥15 million

11 Please write the name of the city and town and number of years that you are staying here.

________ City________Town ___________Years

12 Please tell us about your volunteer activities:

a) Did you participated any kind of environmental volunteer activities during the last one year?

①Did not participated ②Participated (About____days in one year)

b. Did you made any monetary contribution to any environmental volunteer activities during the last one year?

① Did not contributed ② Contributed (About ¥______in one year)

13 Feel free to write any comments or opinion in the following box, that you want to share regarding the Isahaya Bay wetland or Isahaya Bay Reclamation project or anything related to this research survey:

Appendix B Time Value of Money for Environmental Valuation

Cost-benefit analysis of environmental goods often involves decisions to choose among alternatives. The decision to initiate new project involving environment requires using capital allocating or capital budgeting techniques. In conducting cost-benefit analysis of environmental goods or amenities we have to determine whether future benefit of the project is adequately large to defend the current outlays. In order to compare future benefits with the current cost of the project, we must apply the time value of money concept. Because a particular amount of money or benefits received today is worth more than the same amount of money or benefits to be received in future. This is true in any country where investment yields positive rate of return. Thus, this concept of higher comparative present worth of money named *time value of money* becomes very important in all financing decisions and financing projects involving environment are no exception to this. If we attempt to include the opinion of the interest groups in cost-benefit analysis of projects involving environmental goods, we must ensure that all the benefits and costs are adjusted to represent their present worth and accordingly need to apply the time value of money concept.

This appendix considers some of the time value of money concept formulas useful for cost-benefit analysis of environmental goods and are summarized through Table B.1 using the following notations:

FV = Future value, the future worth of an investment after a certain period of time
PV = Present value, the present worth of an investment
A = Annuity, a fixed sum paid or received for a specified number of period
i = Interest rate, discount rate
n = Number of periods
r = Rate of return on an investment
FV_A = Future value of an annuity
PV_A = Present value of an annuity
PV_P = Present value of an annuity in perpetuity
NPV = Net present value
PV_B = Present value of the benefits
PVC = Present value of the costs
PVB = Present value of the benefits

Table B.1. Time value of money formulas and their use in environmental valuation

To find	Formulas	When used
Future value of a single amount	$FV = PV(1+i)^n$	To calculate the future worth of a project investment
Present value of a single amount	$PV = \dfrac{FV_n}{(1+i)^n}$	To calculate the present worth of an amount of benefits to be received in future
Future value of an annuity	$FV_A = A\left[\dfrac{(1+i)^n - 1}{i}\right]$	To calculate the future worth of a project investment to be made through a series of equal payments
Present value of an annuity	$PV_A = A\left[\dfrac{1 - \dfrac{1}{(1+i)^n}}{i}\right]$	To calculate the present worth of benefits to be received in future through a series of equal payments
Net present value	$NPV = PVB - PVC$	To calculate whether the present value of the benefits exceeds the present value of the costs
Present value in perpetuity	$PV_p = \dfrac{A}{i}$	To calculate the present worth of equal benefits to be received in future in perpetuity
Return of a project in perpetuity	$r = \dfrac{A}{PV}$	To calculate the rate of return of a project providing equal benefits to be received in future in perpetuity

Further Readings

For a detailed discussion of the time value of money concept, see Myers 1991; Block and Hirt 1996; Weston and Brigham 1984; Van Horne 1980.

References

Ahmed SU, Gotoh K, Kojima H (2002a) Willingness to pay for restoring the Isahaya Bay Wetland: evidence from contingent valuation method. The annual meeting of the Western Branch of JSCE II-61: B122-123

Ahmed SU, Gotoh K, Kojima H (2002b) Measuring the value of the Isahaya Bay Wetland: an empirical study based on contingent valuation method. 7th Joint Symp. on Science and Technology, Nagasaki University, Session F: Environment:147-150

Ahmed SU, Kojima H, Gotoh K (2002c) Cost-benefit analysis of the Isahaya Bay reclamation project by applying contingent valuation method. Papers on Environmental Information Science 16:305-310

Ahmed SU, Gotoh K, Kojima H (2003) Estimating the willingness to pay for restoring the Isahaya Bay Wetland: evidence from contingent valuation method. Journal of Environmental Systems and Engineering 734/VII-27: 157-165

Amemiya T (1981) Qualitative response models: a survey. Journal of Economic Literature 19: 481-536

Arrow K, Solow R, Portney P, Leamer E, Radner R, Schuman H (1993) Report of the NOAA Panel on contingent valuation. Federal Register 58-10, p4608

Asahi Shimbun (2001) Various Issues 2001-2003

Asiaweek (2001) Public works time bombs, Vol.27, No.6, Feb. 16, 2001

Bateman IJ, Langford IH, Rasbash J (1999) Willingness-to-pay question format in contingent valuation studies. In: Bateman IJ and Willis KG (eds) Valuing environmental preferences: theory and practice of the contingent valuation method in the US, EU, and the developing countries. Oxford University Press, New York. pp 511-539

Bateman IJ, and Willis GK (1999) Introduction and overview. In: Bateman IJ and Willis KG (eds) Valuing environmental preferences- theory and practice of the contingent valuation method in the US, EU, and Developing Countries. Oxford University Press, New York, pp 1-14

Block SB, Hirt GA (1996) Foundation of financial management. Eight edition, Richard D. Irwin

Boardman AE, Greenberg DH, Vining AR and Weimer DL (2001) Cost-benefit analysis: concepts and practice. Second edition, Prentice Hall, New Jersey

Boom P (1972) Estimating demand for public goods: an experiment. European Economic Review 3: 111-130

Cambridge Economics, Inc. (1992) Contingent valuation: a critical assessment. Proceedings of a symposium held in Washington, 2-3 April 1992, Cambridge Economics, Inc., Cambridge, Mass

Cameron TA (1988) A new paradigm for valuing non-market goods using referendum data: maximum likelihood estimation by censored logistic regression. Journal of Environmental Economics and Management. 15:335-379

Carson RT, Wright JL, Carson NJ, Alberini A, and Flores NE (1995) A bibliography of contingent valuation studies and papers. Natural Resource Damage Assessment Inc., San Diego

Costanza R, Farber SC, Maxwell J (1989) Valuation and management of wetland ecosystems, Ecological Economics 1: 334-351

Costanza R, d'Arge R, deGroot R, Farber S, Grasso M, Hannon B, Limburg K, Naeem S, O'Neill RV, Paruelo J, Raskin RG, Sutton PC, van den Belt M (1997) The value of the world's ecosystem services and natural capital. Nature 387: 253-260

de Groot, RS (1992) Functions of nature: evaluation of nature in environmental planning, management and decision making, Groningen, The Netherlands: Wolters Noordhoff BV

Demaris A (1992) Logit Modeling. Sage University paper series on quantitative applications in the social sciences, pp 07-086

Desvousges WH, Smith VK, McGivency MP (1983) A comparison of alternative approaches for estimating recreation and related benefits of water quality improvements. Office of Policy Analysis. US Environmental Protection Agency, Washington

Estrella A (1998) A new measure of fit for equations with dichotomous dependent variables. Journal of Business and Economic Statistics 16(2): 198-205

Food and Agriculture Organization (FAO) (2000) Applications of the contingent valuation method in developing countries- a survey 2000, Economic and Social Development Paper No.146, Editorial Group, FAO Information Division, pp 1-40

Gosselink JG, Odum EP, Pope RM (1974) The value of the tidal marsh. Center for Wetland Resources LSU-SG-74-03, Louisiana State Univ., Baton Rouge

Gotoh K, Ahmed SU, Kojima H (2002) Environmental evaluation of Isahaya Bay Wetland: preliminary tests for designing appropriate CVM questionnaire. Journal of JSCE 87:71-74

Gotoh K, Ahmed SU, Toshine Y (2003) Estimation of the willingness to pay for preserving public parks in Nagasaki City by using contingent valuation method". (Under review in the Journal of Infrastructure Planning and Management, JSCE)

Greene WH (1997) Econometric Analysis. Third Edition. Prentice Hall

Hair JF, Anderson RE, Tatham RL, Black WC (1998) Multivariate data analysis, Prentice-Hall Inc., New Jersey

Hair JD (1988) The economics of conserving wetlands: a widening circle. Paper presented at Workshop on Economics, IUCN General Assembly, February 4-5, Costa Rica

Hanemann WM, Loonis J, Kanninen B (1991) Statistical efficiency of double-bounded dichotomous choice contingent valuation. American Journal of Agricultural Economics 73:1255-1263

Hanemann WM (1992) Pricing in European environment, Scandinavian University Press, Oslo

Hannon B (1993) Sense of place: geographic discounting by people, animals, and plants. Ecological Economics 10:147-173

Hausman JA (1993) Contingent valuation: a critical assessment, Elsevier Science Publishers, BV, Amsterdam

Hoen, JP, Randall A (1987) A satisfactory benefit-cost indicator from contingent valuation. Journal of Environmental Economics and Management 14:226-247

Hosmer D, and Stanley L (1989) Applied logistic regression. Wiley & Sons, New York

Huis Ten Bosch (HTB) (2002) Huis Ten Bosch environmental accounting report: 1992-2001. Environmental Group, Huis Ten Bosch Ltd.

Japan Wetlands Action Network (JWAN) (1999) National Report, May Issue.
HP: http://www.kt.rim.or.jp/~hira/jawan/index-j.html

Kahneman D, Solvic P, Tversky A (1982) Judgement under uncertainty: heuristics and biases. Cambridge University Press, New York

King MD, Mozzata M (2002) Ecosystem valuation. US Department of Agriculture Natural Resources Conservation Service and National Oceanographic and Atmospheric Administration

Kuriyama K (1998) Measuring the value of the ecosystem in the Kushiro Wetland: an empirical study of choice experiments. Forest Economics and Policy Working Paper #9802, Department of Forest Science, Faculty of Agriculture, Hokkaido University

Lesser JA, Dodds DE and Zerbe RO (1997) Environmental economics and policy. Addison-Wisley, USA

Levine DM, Stephan D, Krehbiel TC and Berenson ML (2002) Statistics for managers using Microsoft Excel. Third edition, Prentice Hall, New Jersey.

Loomis JB (1988) Contingent valuation using dichotomous choice models. Journal of Leisure Research 20: 46-56

Loomis J, Hanemann WM, Kanninen B, Wegge T (1991) Willingness to pay to protect wetlands and reduce wildlife contamination from agricultural drainage, The Economics and Management of Water and Draiange in Agriculture. Kluwer Academic Publishers, Boston

McFadden D, and Leonard GK (1993) Issues in the contingent valuation of environmental goods: methodologies for data collection and analysis. In: Hausman JA (eds) Contingent valuation: a critical assessment, North-Holland, New York

Madalla GS (1983) Limited dependent and qualitative variables in econometrics. Cambridge Univ. Press

Mainichi Daily News (2001a) Kyushu fishermen storm land reclamation site, Feb. 25, 2001

..........(2001b) Government set to damn contentious project, Aug. 26, 2001

..........(2001c) Government halts reclamation project, Mar. 7, 2001

.... (2002) Seaweed growers block workers from reclaiming Isahaya Bay, Aug. 1, 2002

Maki AW (1991) The Exxon oil spill: initial environmental impact assessment, Environmental Science and Technology 25: 24-29

Ministry of Agriculture, Fisheries and Forestry (MAFF) (1987) National reclamation project, Kyushu Regional, Agricultural Administration Office, pp 28-34

.... (1990) Message on the Isahaya Bay reclamation project. Agricultural Structure Improvement Bureau, Japan

Ministry of Public Management, Home Affairs, Posts and Telecommunication (MPHP) (2002) Statistical yearbook. Statistical Survey Department, Statistics Bureau, Japan

.........(2003) Statistical yearbook. Statistical Survey Department, Statistics Bureau, Japan

Mitchell CR, Carson RT (1989) Using surveys to value public goods: The contingent valuation method, Resources for the Future, Washington, D.C.

Miyairi K (1998) Public interests of the Isahaya-Bay sea reclamation project and its cost-benefit evaluation. Journal of Business and Economics 77-4:123-161

Myers SC, Brealey RA (1991) Principles of corporate finance. Fourth edition, McGraw-Hill, Inc.

Nature Conservation Society of Japan (NACSJ) (1998) The present status of tidal flats in Japan, Higata Database, March 16, 1998

Neter J, Kutner M, Nachtsheim C, Wasserman W (1996) Applied linear statistical models. Fourth edition, Irwin, Homewood, IL

Nishi Nippon Shinbun (NNS) (2003) Thirteen targets for public works set by the Ministry of Land, Infrastructure and Transport, Japan, May 5, 2003

Nishi-Nippon Telephone and Telegraph Ltd. (NTT) (2002) Hello page: Nagasaki and Sasebo. NTT Quorisu, Tokyo

Pate J, and Loomis J (1997) The effect of distance on willingness to pay values: a case study of wetlands and salmon in California. Ecological Economics 20:199-207

Raftery AE (1995) Bayesian model selection in social research. In: Marsden PV (eds) Sociological Methodology 1995, London

Rice JC (1994) Logistic regression: an introduction. In: Thompson B, (eds) Advances in social science methodology 3, Greenwich, CT: JAI Press, pp 191-245

Snee RD (1973) Some aspects of nonorthogonal data analysis, Part I. Developing prediction equations, Journal of Quality Technology 5: 67-79

Spash CL (2000) Ecosystems, contingent valuation and ethics: the case of wetland recreation. Ecological Economics 34: 195-215

Stynes DJ, George LP (1984) A Review of logit Models with implications for modeling recreation choices. Journal of Leisure Research 16: 295-310

Sutherland RJ, Walsh R (1985) Effect of distance on the preservation value of water quality. Land Economics 67: 390-300

Tabachnick BG, Fidell LS (1996) Using multivariate statistics, Third edition. Harper Collins, New York

The Daily Yomiuri (2003a) Huis Ten Bosch goes belly-up, Feb. 27, 2003, p18

……… (2003b) Outlook grim for Dutch village theme park, Feb 28, 2003, p 16

Thibodeau FR, Ostro BD (1981) An economic analysis of wetland protection. Journal of Environmental Management12: 19-30

Tietenberg T (1996) Environmental and natural resource economics. Harper Collins College Publishers, New York

Toyo Keizai Inc. (2003) Urban city data bank, Toyo Keizai Supplement 137, Tokyo

Turnbull BW (1976) The empirical distribution with arbitrarily grouped, censored and truncated data. Journal of the Royal Statistical Society 3 / V-38: 290-295

Van Horne JC (1980) Financial management and policy. Fifth edition, Prentice-Hall, Inc.

Weston JF, Brigham EF (1984) Essentials of managerial finance. Seventh edition, The Dryden Press

Whitehead JC (2000) A practitioner's premier on contingent valuation. Working paper, Department of Economics, East Carolina University

Whittington D, Briscoe J, Mu X, Barron W (1990) Estimating the willingness to pay for water services in developing countries: a case study of the use of contingent valuation surveys in southern Haiti, Economic Development and Cultural Change 38: 293-311

Subject Index